Use R!

Series Editors:
Robert Gentleman   Kurt Hornik   Giovanni Parmigiani

MW00844343

Use R!

G.P. Nason

Wavelet Methods
in Statistics with R

Springer

G.P. Nason
Department of Mathematics
University of Bristol
University Walk
Bristol BS8 1TW
United Kingdom
g.p.nason@bristol.ac.uk

Series Editors:
Robert Gentleman
Program in Computational Biology
Division of Public Health Sciences
Fred Hutchinson Cancer Research Center
1100 Fairview Avenue, N. M2-B876
Seattle, Washington 98109
USA

Kurt Hornik
Department of Statistik and Mathematik
Wirtschaftsuniversität Wien Augasse 2-6
A-1090 Wien
Austria

Giovanni Parmigiani
The Sidney Kimmel Comprehensive
    Cancer Center at Johns Hopkins University
550 North Broadway
Baltimore, MD 21205-2011
USA

ISBN: 978-0-387-75960-9   e-ISBN: 978-0-387-75961-6
DOI: 10.1007/978-0-387-75961-6

Library of Congress Control Number: 2008931048

Printed on acid-free paper

springer.com

To Philippa, Lucy, Suzannah, Mum and Dad.

Preface

When Zhou Enlai, Premier of the People's Republic of China (1949–1976), was asked his opinion of the French Revolution (1789–1799) he replied "It's too early to tell", see Rosenberg (1999). I believe that the same can be said about wavelets. Although particular wavelets were discovered many years ago, the substantial body of literature that we might today call 'wavelet theory' began to be established during the 1980s. Wavelets were introduced into statistics during the late 1980s and early 1990s, and they were initially popular in the curve estimation literature. From there they spread in different ways to many areas such as survival analysis, statistical time series analysis, statistical image processing, inverse problems, and variance stabilization.

The French Revolution was also the historical backdrop for the introduction of Fourier series which itself raised considerable objections from the scientific establishment of the day, see Westheimer (2001). Despite those early objections, we find that, 200 years later, many new Fourier techniques are regularly being invented in many different fields. Wavelets are also a true scientific revolution. Some of their interesting features are easy to appreciate: e.g., multiscale, localization, or speed. Other important aspects, such as the unconditional basis property, deserve to be better known. I hope that this book, in some small way, enables the creation of many new wavelet methods. Wavelet methods will be developed and important for another 200 years!

This book is about the role of wavelet methods in statistics. My aim is to cover the main areas in statistics where wavelets have found a use or have potential. Another aim is the promotion of the *use* of wavelet methods as well as their *description*. Hence, the book is centred around the freeware R and WaveThresh software packages, which will enable readers to learn about statistical wavelet methods, *use* them, and *modify* them for their own use. Hence, this book is like a traditional monograph in that it attempts to cover a wide range of techniques, but, necessarily, the coverage is biased towards areas that I and WaveThresh have been involved in. A feature is that the code for nearly all the figures in this book is available from the WaveThresh website. Hence, I hope that this book (at least) partially meets the criteria of 'reproducible research' as promoted by Buckheit and Donoho (1995).

Most of `WaveThresh` was written by me. However, many people contributed significant amounts of code and have generously agreed for this to be distributed within `WaveThresh`. I would like to thank Felix Abramovich (FDR thresholding), Stuart Barber (complex-valued wavelets and thresholding, Bayesian wavelet credible interval), Tim Downie (multiple wavelets), Idris Eckley (2D locally stationary wavelet processes), Piotr Fryzlewicz (Haar–Fisz transform for Poisson), Arne Kovac (wavelet shrinkage for irregular data), Todd Ogden (change-point thresholding), Theofanis Sapatinas (Donoho and Johnstone test functions, some wavelet packet time series code, BayesThresh thresholding), Bernard Silverman (real FFT), David Herrick (wavelet density estimation), and Brani Vidakovic (Daubechies-Lagarias algorithm). Many other people have written add-ons, improvements, and extensions, and these are mentioned in the text where they occur. I would like to thank Anthony Davison for supplying his group's SBand code.

I am grateful to A. Black and D. Moshal of the Dept. of Anaesthesia, Bristol University for supplying the plethysmography data, to P. Fleming, A. Sawczenko, and J. Young of the Bristol Institute of Child Health for supplying the infant ECG/sleep state data, to the Montserrat Volcano Observatory and Willy Aspinall, of Aspinall and Associates, for the RSAM data.

Thanks to John Kimmel of Springer for encouraging me to write for the Springer UseR! series. I have had the pleasure of working and interacting with many great people in the worlds of wavelets, mathematics, and statistics. Consequently, I would like to thank Felix Abramovich, Anestis Antoniadis, Dan Bailey*, Rich Baraniuk, Stuart Barber, Jeremy Burn, Alessandro Cardinali, Nikki Carlton, Merlise Clyde, Veronique Delouille, David Donoho, Tim Downie, Idris Eckley, Piotr Fryzlewicz*, Gérard Grégoire, Peter Green, Peter Hall, David Herrick, Katherine Hunt, Maarten Jansen, Iain Johnstone, Eric Kolaczyk, Marina Knight*, Gerald Kroisandt, Thomas Lee, Emma McCoy, David Merritt, Robert Morgan, Makis Motakis, Mahadevan Naventhan, Matt Nunes*, Sofia Olhede, Hee-Seok Oh, Marianna Pensky, Howell Peregrine, Don Percival, Marc Raimondo, Theofanis Sapatinas, Sylvain Sardy, Andrew Sawczenko, Robin Sibson, Glenn Stone, Suhasini Subba Rao, Kostas Triantafyllopoulos, Brani Vidakovic, Sebastien Van Bellegem, Rainer von Sachs, Andrew Walden, Xue Wang, Brandon Whitcher. Those marked with * in the list are due special thanks for reading through large parts of the draft and making a host of helpful suggestions. Particular thanks to Bernard Silverman for introducing me to wavelets and providing wise counsel during the early stages of my career.

Bristol,   *Guy Nason*
                                                                        March 2006

Contents

1

Introduction

1.1 What Are Wavelets?

This section is a highlight of the next chapter of this book, which provides an in-depth introduction to wavelets, their properties, how they are derived, and how they are used.

Wavelets, as the name suggests, are 'little waves'. The term 'wavelets' itself was coined in the geophysics literature by Morlet et al. (1982), see Daubechies (1992, p. vii). However, the evolution of wavelets occurred over a significant time scale and in *many* disciplines (including statistics, see Chapter 2). In later chapters, this book will explain some of the key developments in wavelets and wavelet theory, but it is not a comprehensive treatise on the fascinating history of wavelets. The book by Heil and Walnut (2006) comprehensively covers the early development of wavelets. Many other books and articles contain nice historical descriptions including, but not limited to, Daubechies (1992), Meyer (1993a), and Vidakovic (1999a).

Since wavelets, and wavelet-like quantities have turned up in many disciplines it is difficult to know where to begin describing them. For example, if we decided to describe the Fourier transform or Fourier series, then it would be customary to start off by defining the Fourier basis functions $(2\pi)^{-1/2}e^{inx}$ for integers n. Since this is a book on 'wavelets in statistics', we could write about the initial developments of wavelets in statistics in the early 1990s that utilized a particular class of wavelet transforms. Alternatively, we could start the story from a signal processing perspective during the early to mid-1980s, or earlier developments still in mathematics or physics. In fact, we begin at a popular starting point: the *Haar wavelet*. The Haar *mother* wavelet is a mathematical function defined by

$$\psi(x) = \begin{cases} 1 & x \in [0, \frac{1}{2}), \\ -1 & x \in [\frac{1}{2}, 1), \\ 0 & \text{otherwise,} \end{cases} \tag{1.1}$$

and it forms the basis of our detailed description of wavelets in Chapter 2. The Haar wavelet is a good choice for educational purposes as it is very simple, but it also exhibits many characteristic features of wavelets. Two relevant characteristics are the oscillation (the Haar wavelet 'goes up and down'; more mathematically this can be expressed by the condition that $\int_{-\infty}^{\infty} \psi(x)\,dx = 0$, a property shared by all wavelets) and the compact support (not all wavelets have compact support, but they must decay to zero rapidly). Hence, wavelets are objects that oscillate but decay fast, and hence are 'little'.

Once one has a mother wavelet, one can then generate *wavelets* by the operations of dilation and translation as follows. For integers j, k we can form

$$\psi_{j,k}(x) = 2^{j/2}\psi(2^j x - k). \tag{1.2}$$

It turns out (again see Chapter 2) that such wavelets can form an orthonormal set. In other words:

$$< \psi_{j,k}, \psi_{j',k'} >= \int_{-\infty}^{\infty} \psi_{j,k}(x)\psi_{j',k'}(x)\,dx = \delta_{j,j'}\delta_{k,k'}, \tag{1.3}$$

where $\delta_{m,n} = 1$ if $m = n$, and $\delta_{m,n} = 0$ if $m \neq n$. Here $< \cdot, \cdot >$ is the inner product, see Section B.1.3. Moreover, such a set of wavelets can form bases for various spaces of functions. For example, and more technically, $\{\psi_{j,k}(x)\}_{j,k\in\mathbb{Z}}$ can be a complete orthonormal basis for $L^2(\mathbb{R})$, see Walter and Shen (2005, p. 10). So, given a function $f(x)$, we can decompose it into the following generalized Fourier series as

$$f(x) = \sum_{j=-\infty}^{\infty} \sum_{k=-\infty}^{\infty} d_{j,k}\psi_{j,k}(x), \tag{1.4}$$

where, due to the orthogonality of the wavelets, we have

$$d_{j,k} = \int_{-\infty}^{\infty} f(x)\psi_{j,k}(x)\,dx =< f, \psi_{j,k} >, \tag{1.5}$$

for integers j, k. The numbers $\{d_{j,k}\}_{j,k\in\mathbb{Z}}$ are called the *wavelet coefficients* of f.

Although we have presented the above equations with the Haar wavelet in mind, they are equally valid for a wide range of other wavelets, many of which are described more fully in Chapter 2. Many 'alternative' wavelets are more appropriate for certain purposes mainly because they are smoother than the discontinuous Haar wavelet (and hence they also have better decay properties in the Fourier domain as well as the time domain).

1.2 Why Use Wavelets?

Why use wavelets? This is a good 'frequently asked question'. There are good reasons why wavelets can be useful. We outline the main reasons in this section and amplify on them in later sections. The other point to make is that wavelets are not a panacea. For many problems, wavelets are effective, but there are plenty of examples where existing methods perform just as well or better. Having said that, in many situations, wavelets often offer a kind of insurance: they will *sometimes* work better than certain competitors on *some* classes of problems, but *typically* work nearly as well on *all* classes. For example, one-dimensional (1D) nonparametric regression has mathematical results of this type. Let us now describe some of the important properties of wavelets.

Structure extraction. Equation (1.5) shows how to compute the wavelet coefficients of a function. Another way of viewing Equation (1.5) is to use the inner product notation, and see that $d_{j,k}$ quantifies the 'amount' of $\psi_{j,k}(x)$ that is 'contained' within $f(x)$. So, if the coefficient $d_{j,k}$ is large, then this means that there is some oscillatory variation in $f(x)$ near $2^{-j}k$ (assuming the wavelet is localized near 0) with an oscillatory wavelength proportional to 2^{-j}.

Localization. If $f(x)$ has a discontinuity, then this will only influence the $\psi_{j,k}(x)$ that are near it. Only those coefficients $d_{j,k}$ whose associated wavelet $\psi_{j,k}(x)$ overlaps the discontinuity will be influenced. For example, for Haar wavelets, the only Haar coefficients $d_{j,k}$ that can possibly be influenced by a discontinuity at x^* are those for which j, k satisfy $2^{-j}k \leq x^* \leq 2^{-j}(k + 1)$. For the Haar wavelets, which do not themselves overlap, only one wavelet per scale overlaps with a discontinuity (or other feature). This property is in contrast to, say, the Fourier basis consisting of sine and cosine functions at different frequencies: *every* basis sine/cosine will interact with a discontinuity no matter where it is located, hence influencing every Fourier coefficient. Both of the properties mentioned above can be observed in the image displayed in Figure 1.1. The original image in the top left of Figure 1.1 contains many edges which can be thought of as discontinuities, i.e., sharp transitions where the grey level of the image changes rapidly. An image is a two-dimensional (2D) object, and the wavelet coefficients here are themselves 2D at different scales (essentially the k above changes from being 1D to 2D).

The edges are clearly apparent in the wavelet coefficient images, particularly at the fine and medium scales, and these occur very close to the positions of the corresponding edges in the original image. The edge of the teddy's head can also be seen in the coarsest scale coefficients. What about wavelets being contained within an image? In the top right subimage containing the fine-scale coefficients in Figure 1.1, one can clearly see the chequered pattern of the tablecloth. This pattern indicates that the width of the squares is similar to the wavelength of the wavelets generating the coefficients. Figure 1.1 showed the values of the wavelet coefficients. Figure 1.2 shows the approximation possible by using all wavelets (multiplied by their respective coefficients) up to and including a particular scale. Mathematically, this can be represented by the following formula, which is a restriction of Formula (1.4):

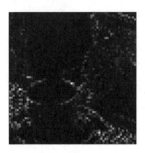

Fig. 1.1. *Top left:* teddy image. Wavelet transform coefficients of teddy at a selection of scales: fine scale *(top right),* medium scale *(bottom left),* and coarse scale *(bottom right).*

$$f_J(x) = \sum_{j=-\infty}^{J} \sum_{k=-\infty}^{\infty} d_{j,k}\psi_{j,k}(x). \tag{1.6}$$

In Figure 1.2, the top right figure contains the finest wavelets and corresponds to a larger value of J than the bottom right figure. The overall impression is that the top right figure provides a fine-scale approximation of the original image, the bottom left an intermediate scale approximation, whereas the bottom right image is a very coarse representation of the original.

Figure 2.26 on page 29 shows another example of a 1D signal being approximated by a Haar wavelet representation at different scales.

Figures 1.1, 1.2, and 2.26 highlight how wavelets can separate out information at different scales and provide localized information about that activity. The pictures provide 'time-scale' information.

Efficiency. Figure 1.3 provides some empirical information about execution times of both a wavelet transform (`wd` in `WaveThresh`) and the fast Fourier transform (`fft` in `R`). The figure was produced by computing the two transforms on data sets of size n (for various values of n) repeating those computations many times, and obtaining average execution times. Figure 1.3 shows

Fig. 1.2. *Top left:* teddy image. Wavelet approximation of teddy at fine scale *(top right)*, medium scale *(bottom left)*, and coarse scale *(bottom right)*.

the average execution times divided by n for various values of n. Clearly, the execution time for wavelets (divided by n) looks roughly constant. Hence, the computation time for the wavelet transformation itself should be proportional to n. However, the execution time (divided by n) for the fft is still increasing as a function of n. We shall see theoretically in Chapter 2 that the computational effort of the (basic) discrete wavelet transform is of order n compared to order $n \log n$ for the fft.

From these results, one can say that the wavelet transform is faster (in terms of order) than the fast Fourier transform. However, we need to be careful since (i) the two transforms perform different jobs and (ii) actually, from Figure 1.3 it appears that the fast Fourier transform is faster than the wavelet one for $n \leq 125000$ (although this latter statement is highly dependent on the computing environment). However, it is clear that the wavelet transform *is* a fast algorithm. We shall later also learn that it is also just as efficient in terms of memory usage.

Sparsity. The next two plots exhibit the sparse nature of wavelet transforms for many real-life functions. Figure 1.4 (top) shows a picture of a simple piecewise polynomial that originally appeared in Nason and Silverman (1994). The specification of this polynomial is

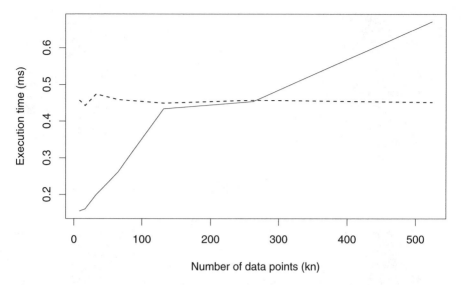

Fig. 1.3. Average execution times (divided by n) of R implementation of fast Fourier transform *(solid line),* and wavelet transform wd *(dashed line).* The *horizontal axis* is calibrated in thousands of n, i.e., so 500 corresponds to $n = 500000$.

$$y(x) = \begin{cases} 4x^2(3 - 4x) & \text{for } x \in [0, 1/2), \\ \frac{4}{3}x(4x^2 - 10x + 7) - \frac{3}{2} & \text{for } x \in [1/2, 3/4), \\ \frac{16}{3}x(x - 1)^2 & \text{for } x \in [3/4, 1]. \end{cases} \quad (1.7)$$

The precise specification is not that important, but it is essentially three cubic pieces that join at $3/4$, and a jump at $1/2$. Figure 1.4 (bottom) shows the wavelet coefficients of the piecewise polynomial. Each coefficient is depicted by a small vertical line. The coefficients $d_{j,k}$ corresponding to the same resolution level j are arranged along an imaginary horizontal line. For example, the finest-resolution-level coefficients corresponding to $j = 8$ appear as the lowest set of coefficients arranged horizontally in the bottom plot of Figure 1.4. Coefficients with $2^{-j}k$ near zero appear to the left of the plot, and near one to the right of the plot. Indeed, one can see that the coefficients are closer together at the finer scales; this is because 2^{-j} is smaller for larger j.

There are few non-zero coefficients in Figure 1.4. Indeed, a rough count of these shows that there appears to be about 10 non-zero, and approximately one that seems 'big'. So, the 511 non-zero samples of the piecewise polynomial, of which about 90% are greater in size than 0.2, are transformed into about 10 non-zero wavelet coefficients (and many of the remaining ones are not just very small but actually zero).

It is not easy to see the pattern of coefficients in Figure 1.4. This is because the coefficients are all plotted to the same vertical scale, and there is only one really large coefficient at resolution level zero, and all the others are relatively smaller. Figure 1.5 (bottom) shows the same coefficients but plotted so that

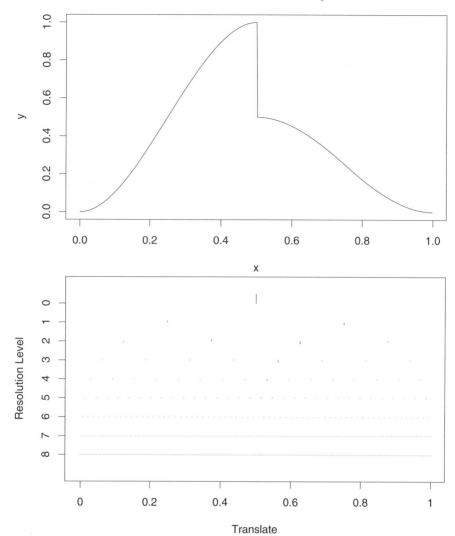

Fig. 1.4. *Top:* piecewise polynomial function sampled at 512 equally spaced locations in [0, 1] (reproduced with permission from Nason and Silverman (1994)). *Bottom:* wavelet coefficients of piecewise polynomial function. All coefficients plotted to same scale.

each resolution level of coefficients is independently scaled (so here the medium to finer-scale coefficents have been scaled up so that they can be seen). One can see that the significant coefficients in the bottom plot 'line up' with the discontinuity in the piecewise polynomial. This is an illustration of the comment above that wavelet coefficients can be large when their underlying corresponding wavelets overlap the 'feature of interest' such as discontinuities. Another way of thinking about this is to view the discontinuity in the top plot of Figure 1.4 as an edge, and then see that the wavelet coefficients are clustered around the edge location (much in the same way as the image wavelet coefficients in Figure 1.1 cluster around corresponding edge locations in the original image). Figures 2.6 and 2.7 show similar sets of wavelet coefficient plots for two different functions. The two original functions are more complex than the piecewise polynomial, but the comments above about sparsity and localization still apply.

The sparsity property of wavelets depends on the (deep) mathematical fact that wavelets are *unconditional bases* for many function spaces. Indeed, Donoho (1993b) notes "an orthogonal basis which is an unconditional basis for a function class $\mathcal{F}$ is better than other orthogonal bases in representing elements of $\mathcal{F}$, because it typically compresses the energy into a smaller number of coefficients". Wavelet series offer unconditional convergence, which means that partial sums of wavelet series converge irrespective of the order in which the terms are taken. This property permits procedures such as forming well-defined estimates by accumulating wavelet terms in (absolute) size order of the associated wavelet coefficients. This is something that cannot always be achieved with other bases, such as Fourier, for certain important and relevant function spaces. More information can be found in Donoho (1993b), Hazewinkel (2002), and Walker (2004).

Not sparse! Taking the wavelet transform of a sequence does not always result in a sparse set of wavelet coefficients. Figure 1.6 shows the wavelet transform coefficients of a sequence of 128 independent standard normal random variates. The plot does not suggest a sparsity in representation. If anything the coefficients appear to be 'spread out' and fairly evenly distributed. Since the wavelet transform we used here is an orthogonal transformation, the set of coefficients also forms an iid Gaussian set. Hence, the distribution of the input variates is invariant to the wavelet transformation and no 'compression' has taken place, in contrast to the deterministic functions mentioned above. Later, we shall also see that the wavelet transform conserves 'energy'. So, the wavelet transform can squeeze a signal into fewer, often larger, coefficients, but the noise remains 'uncompressed'. Hence taking the wavelet transform often dramatically improves the signal-to-noise ratio.

For example, the top plot in Figure 1.5 shows that the piecewise polynomial is a function with values between zero and one. The 'energy' (sum of the squared values of the function) is about 119.4. The 'energy' of the wavelet coefficients is the same, but, as noted above, many of the values are zero, or very close to zero. Hence, since energy is conserved, and many coefficients

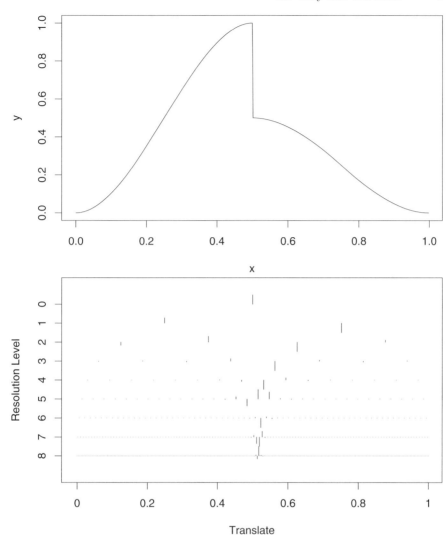

Fig. 1.5. *Top:* piecewise polynomial function (again) sampled at 512 equally spaced locations in [0, 1] (reproduced with permission from Nason and Silverman (1994)). *Bottom:* wavelet coefficients of piecewise polynomial function. Each horizontal-scale level of coefficients has been scaled separately to make the coefficient of largest absolute size in each row the same apparent size in the plot.

were smaller than in the original function, some of them must be larger. This is indeed the case: the largest coefficient is approximately 6.3, and the other few large coefficients are above one. Thus, if we added noise to the input, the output would have a higher signal-to-noise ratio, just by taking the wavelet transform.

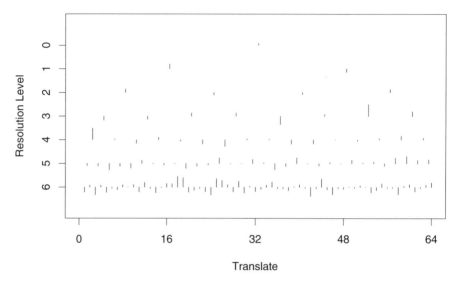

Fig. 1.6. Wavelet transform coefficients of a sequence of 128 independent standard normal variates.

Efficiency (again). Looking again at the bottom plots in Figures 1.4 and 1.5 one can see that there is one coefficient at resolution level zero, two at level one, four at level two, and generally 2^j at level j. The method we used is known as a pyramid algorithm because of the pyramidal organization of the coefficients. The algorithm for general wavelets is the discrete wavelet transform due to Mallat (1989b). The total number of coefficients shown in each of these plots is

$$1 + 2 + 4 + 8 + 16 + 32 + 64 + 128 + 256 = 511,$$

and actually there is another coefficient that is not displayed (but that we will learn about in Chapter 2). This means there are 512 coefficients in total and the same number of coefficients as there were samples from the original function. As we will see later, the pyramid algorithm requires only a fixed number of computations to generate each coefficient. Hence, this is another illustration that the discrete wavelet transform can be computed using order N computational operations.

Summary. The key features of wavelet methods are as follows:

- Sparsity of representation for a wide range of functions including those with discontinuities;
- The ability to 'zoom in' to analyze functions at a number of scales and also to manipulate information at such scales;
- Ability to detect and represent localized features and also to create localized features on synthesis;
- Efficiency in terms of computational speed and storage.

The individual properties, and combinations of them, are the reasons why wavelets are useful for a number of statistical problems. For example, as we will see in Chapters 3 and 4, the sparsity of representation, especially for functions with discontinuities, is extremely useful for curve estimation. This is because the wavelet transform turns the problem from one where a function is estimated at many sample locations (e.g. 512 for the piecewise polynomial) to one where the values of fewer coefficients need to be estimated (e.g. very few for the piecewise polynomial). Thus, the ratio of 'total number of data points' to 'number of things that need to be estimated' is often much larger after wavelet transformation and hence can lead to better performance.

1.3 Why Wavelets in Statistics?

It would take a whole book on its own to catalogue and describe the many applications of wavelets to be found in a wide range of disciplines, so we do not attempt that here. One of the reasons for the impact and diversity of applications for wavelets is that they are, like Fourier series and transforms, highly functional tools with valuable properties, and hence they often end up as the tool of choice in certain applications.

It is our intention to describe, or at least mention, the main uses of wavelets in statistics in the later chapters of this book. Alternative reviews on wavelets and their statistical uses can be found in the papers by Antoniadis (1997), Abramovich et al. (2000), and Antoniadis (2007). Existing books on wavelets and statistics include Ogden (1997), Vidakovic (1999b) on general statistics, Jansen (2001) for noise reduction, Percival and Walden (2000) on wavelets and time series analysis, and Gencay et al. (2001), which treats wavelets and some extensions and their application through filtering to stationary time series, wavelet denoising and artificial neural networks. Naturally, much useful review material appears in many scientific papers that are referenced throughout this book.

Chapter 2 of this book provides a general introduction to wavelets. It first introduces the Haar wavelet transform by looking at successive pairwise differencing and aggregation and then moves on to more general, smoother, wavelets. The chapter examines the important properties of wavelets in more detail and then moves on to some important extensions of the basic discrete wavelet transform.

Chapters 3 to 6 examine three statistical areas where wavelets have been found to be useful. Chapter 3 examines the many methods that use wavelets for estimation in nonparametric regression problems for equally spaced data with Gaussian iid noise. Wavelets are useful for such nonparametric problems because they form sparse representations of functions, including those with discontinuities or other forms of inhomogeneity. Chapter 4 then looks at important variations for data that are correlated, non-Gaussian, and not necessarily equally spaced. The chapter also addresses the question of confidence intervals for wavelet estimates and examines wavelet methods for density estimation, survival, and hazard rate estimation and the solution of inverse problems. Sparsity is also key here.

Chapter 5 considers how wavelets can be of use for both stationary and nonstationary time series analysis. For nonstationary time series the key properties are the wavelet oscillation itself and the ability of wavelets to localize information in time and scale simultaneously. Chapter 6 provides an introduction to how wavelets can be used as effective variance stabilizers, which can be of use in mean estimation for certain kinds of non-Gaussian data. In variance stabilization the key wavelet properties are sparsity (for estimation) and localization (for localized variance stabilization).

The fast and efficient algorithms underlying wavelets benefit all of the areas mentioned above.

If the reader already has a good grasp of the basics of wavelets, then they can safely ignore Chapter 2 and move straight on to the statistical Chapters 3 to 6. On the other hand, if the reader wants to learn the minimum amount about wavelets, they can 'get away with' reading Sections 2.1 to 2.7 inclusive and still be in a position to understand most of the statistical chapters. Also, each of the statistical chapters should be able to be read independently with, perhaps, the exception of Chapter 4, which sometimes relies on discussion to be found in Chapter 3.

The reader may note that the style of this book is not that of a usual research monograph. This difference in style is deliberate. The idea of the book is twofold. One aim is to supply enough information on the background and theory of the various methods so that the reader can understand the basic idea, and the associated advantages and disadvantages. Many readers will be able to obtain full details on many of the techniques described in this book via online access, so there seems little point reproducing them verbatim here. The author hopes that eventually, through various open access initiatives, everybody will be able to rapidly access all source articles.

1.4 Software and This Book

As well as learning about wavelets and their uses in statistics, another key aim of this book is to enable the reader to quickly get started in using wavelet methods via the WaveThresh package in R. The R package, see R Development Core Team (2008), can be obtained from the Comprehensive R Archive Network at cran.r-project.org, as can WaveThresh, which can also be obtained at www.stats.bris.ac.uk/~wavethresh. WaveThresh first became available in 1993 with version 2.2 for the commercial 'version' of R called S-Plus. Since then R has matured significantly, and WaveThresh has increased in size and functionality. Also, many new wavelet-related packages for R have appeared; these are listed and briefly described in Appendix A. Two other excellent packages that address statistical problems are S+Wavelets for the S-PLUS package (see www.insightful.com) and the comprehensive WaveLab package developed for the Matlab package and available from www-stat.stanford.edu/~wavelab. In addition to providing a the list of general wavelet software for R in Appendix A, we will describe other individual specialist software packages throughout the text where appropriate.

Another aim of this book is to provide multiple snippets of R code to illustrate the techniques. Thus, the interested reader with R and WaveThresh installed will be able to reproduce many examples in this book and, importantly, modify the code to suit their own purposes. The current chapter is unusual in this book as it is the only one without detailed R code snippets. All the R code snippets are set in a Courier-like font. The > symbol indicates the R prompt and signifies input commands; code without this indicates R output. The + symbol indicates the R line-continuation symbol when a command is split over multiple lines.

Also available at the WaveThresh website is the code that produced each of the figures. For the code-generated figures we have indicated the name of the function that produced that figure. All these functions are of the form f.xxx(), where xxx indexes the figure within that chapter. So, e.g., f.tsa1() is the first figure available within Chapter 5 on time series analysis.

2

Wavelets

The word 'multiscale' can mean many things. However, in this book we are generally concerned with the representation of objects at a set of scales and then manipulating these representations at several scales simultaneously.

One main aim of this book is to explain the role of wavelet methods *in statistics*, and so the current chapter is necessarily a rather brief introduction to wavelets. More mathematical (and authoritative) accounts can be found in Daubechies (1992), Meyer (1993b), Chui (1997), Mallat (1998), Burrus et al. (1997), and Walter and Shen (2001). A useful article that charts the history of wavelets is Jawerth and Sweldens (1994). The book by Heil and Walnut (2006) contains many important early papers concerning wavelet theory.

Statisticians also have reason to be proud. Yates (1937) introduced a fast computational algorithm for the (hand) analysis of observations taken in a factorial experiment. In modern times, this algorithm might be called a 'generalized FFT', but it is also equivalent to a Haar wavelet packet transform, which we will learn about later in Section 2.11. So statisticians have been 'doing' wavelets, and wavelet packets, since at least 1937!

2.1 Multiscale Transforms

2.1.1 A multiscale analysis of a sequence

Before we attempt formal definitions of wavelets and the wavelet transform we shall provide a gentle introduction to the main ideas of multiscale analysis. The simple description we give next explains the main features of a wavelet transform.

As many problems in statistics arise as a sequence of data observations, we choose to consider the wavelet analysis of sequences rather than functions, although we will examine the wavelet transform of functions later. Another reason is that we want to use R to illustrate our discussion, and R naturally handles discrete sequences (vectors).

We begin with discrete sequence (vector) of data: $y = (y_1, y_2, \ldots, y_n)$, where each y_i is a real number and i is an integer ranging from one to n. For our illustration, we assume that the length of our sequence n is a power of two, $n = 2^J$, for some integer $J \geq 0$. Setting $n = 2^J$ should not be seen as an absolute limitation as the description below can be modified for other n. We call a sequence where $n = 2^J$ a *dyadic* one.

The following description explains how we extract multiscale 'information' from the vector y. The key information we extract is the 'detail' in the sequence at different scales and different locations. Informally, by 'detail' we mean 'degree of difference' or (even more roughly) 'variation' of the observations of the vector at the given scale and location.

The first step in obtaining the detail we require is

$$d_k = y_{2k} - y_{2k-1}, \tag{2.1}$$

for $k = 1, \ldots, n/2$. So, for example, $d_1 = y_2 - y_1$, $d_2 = y_4 - y_3$, and so on. Operation (2.1) extracts 'detail' in that if y_{2k} is very similar to y_{2k-1}, then the coefficient d_k will be very small. If $y_{2k} = y_{2k-1}$ then the d_k will be exactly zero. This seemingly trivial point becomes extremely important later on. If y_{2k} is very different from y_{2k-1}, then the coefficient d_k will be very large.

Hence, the sequence d_k encodes the difference between successive pairs of observations in the original y vector. However, $\{d_k\}_{k=1}^{n/2}$ is *not* the more conventional *first difference vector* (`diff` in R). Specifically, differences such as $y_3 - y_2$ are missing from the $\{d_k\}$ sequence. The $\{d_k\}$ sequence encodes the difference or *detail* at locations (approximately) $(2k + 2k - 1)/2 = 2k - 1/2$.

We mentioned above that we wished to obtain 'detail' at several different scales and locations. Clearly the $\{d_k\}$ sequence gives us information at several different locations. However, each d_k only gives us information about a particular y_{2k} and its *immediate* neighbour. Since there are no closer neighbours, the sequence $\{d_k\}$ gives us information at and around those points y_{2k} at the *finest* possible scale of detail. How can we obtain information at coarser scales? The next step will begin to do this for us.

The next step is extremely similar to the previous one except the subtraction in (2.1) is replaced by a summation:

$$c_k = y_{2k} + y_{2k-1} \tag{2.2}$$

for $k = 1, \ldots, n/2$. This time the sequence $\{c_k\}_{k=1}^{n/2}$ is a set of scaled local averages (scaled because we failed to divide by two, which a proper mean would require), and the information in $\{c_k\}$ is a coarsening of that in the original y vector. Indeed, the operation that turns $\{y_i\}$ into $\{c_k\}$ is similar to a moving average smoothing operation, except, as with the differencing above, we average non-overlapping consecutive pairs. Contrast this to regular moving averages, which average over overlapping consecutive pairs.

An important point to notice is that each c_k contains information originating from both y_{2k} and y_{2k-1}. In other words, it includes information from two adjacent observations. If we now wished to obtain coarser *detail* than contained in $\{d_k\}$, then we could compare two adjacent c_k.

Before we do this, we need to introduce some further notation. We first introduced finest-scale detail d_k. Now we are about to introduce coarser-scale detail. Later, we will go on to introduce detail at successively coarser scales. Hence, we need some way of keeping track of the scale of the detail. We do this by introducing another subscript, j (which some authors represent by a superscript). The original sequence y consisted of 2^J observations. The finest-level detail $\{d_k\}$ consists of $n/2 = 2^{J-1}$ observations, so the extra subscript we choose for the finest-level detail is $j = J - 1$ and we now refer to the d_k as $d_{J-1,k}$. Sometimes the comma is omitted when the identity and context of the coefficients is clear, i.e., $d_{j,k}$. Thus, the finest-level averages, or smooths, c_k are renamed to become $c_{J-1,k}$.

To obtain the next coarsest detail we repeat the operation of (2.1) to the finest-level averages, $c_{J-1,k}$ as follows:

$$d_{J-2,\ell} = c_{J-1,2\ell} - c_{J-1,2\ell-1}, \tag{2.3}$$

this time for $\ell = 1, \ldots n/4$. Again, $d_{J-2,\ell}$ encodes the difference, or detail present, between the coefficients $c_{J-1,2\ell}$ and $c_{J-1,2\ell-1}$ in *exactly the same way* as for the finer-detail coefficient in (2.1). From a quick glance of (2.3) it does not immediately appear that $d_{J-2,\ell}$ is at a different scale from $d_{J-1,k}$. However, writing the $c_{J-1,\cdot}$ in terms of their constituent parts as defined by (2.2), gives

$$d_{J-2,\ell} = (y_{4\ell} + y_{4\ell-1}) - (y_{4\ell-2} + y_{4\ell-3}) \tag{2.4}$$

for the same ℓ as in (2.3). For example, if $\ell = 1$, we have $d_{J-2,1} = (y_4 + y_3) - (y_2 + y_1)$. It should be clear now that $d_{J-2,\ell}$ is a set of differences of components that are averages of two original data points. Hence, they can be thought of as 'scale two' differences, whereas the $d_{J-1,k}$ could be thought of as 'scale one' differences. This is our first encounter with multiscale: we have differences that exist at two different scales.

Scale/level terminology. At this point, we feel the need to issue a warning over terminology. In the literature the words 'scale', 'level', and occasionally 'resolution' are sometimes used interchangeably. In this book, we strive to use 'level' for the integral quantity j and 'scale' is taken to be the quantity 2^j (or 2^{-j}). However, depending on the context, we sometimes use scale to mean level. With the notation in this book j larger (positive) corresponds to finer scales, j smaller to coarser scales.

Now nothing can stop us! We can repeat the averaging Formula (2.2) on the $c_{J-1,k}$ themselves to obtain

$$c_{J-2,\ell} = c_{J-1,2\ell} + c_{J-1,2\ell-1} \tag{2.5}$$

for $\ell = 1, \ldots n/4$. Writing (2.5) in terms of the original vector y for $\ell = 1$ gives $c_{J-2,1} = (y_2 + y_1) + (y_4 + y_3) = y_1 + y_2 + y_3 + y_4$: the local mean of the first four observations without the $\frac{1}{4}$—again $c_{J-2,\ell}$ is a kind of moving average.

By repeating procedures (2.1) and (2.2) we can continue to produce both detail and smoothed coefficients at progressively coarser scales. Note that the actual scale increases by a factor of two each time and the number of coefficients at each scale decreases by a factor of two. The latter point also tells us when the algorithm stops: when only one c coefficient is produced. This happens when there is only $2^0 = 1$ coefficient, and hence this final coefficient must have level index $j = 0$ (and be $c_{0,1}$).

Figure 2.1 shows the (2.1) and (2.2) operations in block diagram form. These kinds of diagrams are used extensively in the literature and are useful for showing the main features of multiscale algorithms. Figure 2.1 shows the generic step of our multiscale algorithm above. Essentially an input vector $c_j = (c_{j,1}, c_{j,2}, \ldots, c_{j,m})$ is transformed into two output vectors c_{j-1} and d_{j-1} by the above mathematical operations. Since Figure 2.1 depicts the 'generic

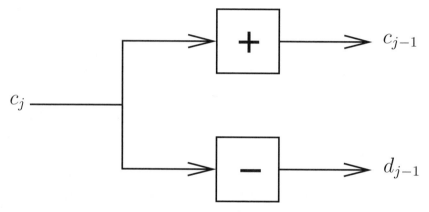

Fig. 2.1. Generic step in 'multiscale transform'. The input *vector*, c_j, is transformed into two output vectors, c_{j-1} and d_{j-1}, by the addition and subtraction operations defined in Equations (2.1) and (2.2) for $j = J, \ldots, 1$.

step', the figure also implicitly indicates that the output c_{j-1} will get fed into an identical copy of the block diagram to produce vectors c_{j-2} and d_{j-2} and so on. Figure 2.1 does not show that the initial input to the 'multiscale algorithm' is the input vector y, although it could be that $c^J = y$. Also, the figure does not clearly indicate that the length of c_{j-1} (and d_{j-1}) is half the length of c_j, and so, in total, *the number of output elements of the step is identical to the number of input elements*.

Example 2.1. Suppose that we begin with the following sequence of numbers: $y = (y_1, \ldots, y_n) = (1, 1, 7, 9, 2, 8, 8, 6)$. Since there are eight elements of y, we have $n = 8$ and hence $J = 3$ since $2^3 = 8$. First apply Formula (2.1) and simply subtract the first number from the second as follows: $d_{2,1} = y_2 - y_1 = 1 - 1 = 0$. For the remaining d coefficients at level $j = 2$ we obtain $d_{2,2} = y_4 - y_3 = 9 - 7 =$

2, $d_{2,3} = y_6 - y_5 = 8 - 2 = 6$ and finally $d_{2,4} = y_8 - y_7 = 6 - 8 = -2$. As promised there are $2^{J-1} = n/2 = 4$ coefficients at level 2.

For the 'local average', we perform the same operations as before but replace the subtraction by addition. Thus, $c_{2,1} = y_2 + y_1 = 1 + 1 = 2$ and for the others $c_{2,2} = 9 + 7 = 16$, $c_{2,3} = 8 + 2 = 10$, and $c_{2,4} = 6 + 8 = 14$.

Notice how we started off with eight y_i and we have produced four $d_{2,\cdot}$ coefficients and four $c_{2,\cdot}$ coefficients. Hence, we produced as many output coefficients as input data. It is useful to write down these computations in a graphical form such as that depicted by Figure 2.2. The organization of

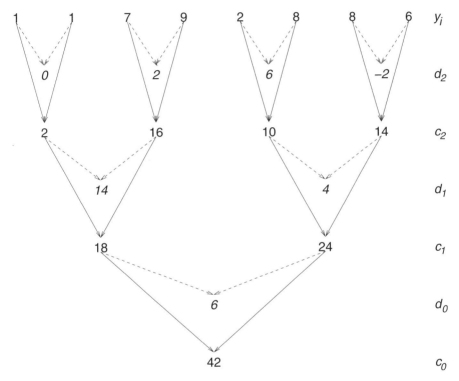

Fig. 2.2. Graphical depiction of a multiscale transform. The *dotted arrows* depict a subtraction and *numbers in italics* the corresponding detail coefficient $d_{j,k}$. The *solid arrows* indicate addition, and *numbers set in the upright font* correspond to the $c_{j,k}$.

coefficients in Figure 2.2 can be visualized as an inverted pyramid (many numbers at the top, one number at the bottom, and steadily decreasing from top to bottom). The algorithm that we described above is an example of a *pyramid* algorithm.

The derived coefficients in Figure 2.2 all provide information about the original sequence in a scale/location fashion. For example, the final 42 indicates that the sum of the *whole* original sequence is 42. The 18 indicates that the sum of the first four elements of the sequence is 18. The *4* indicates that the sum of the last quarter of the data minus the sum of the third quarter is four. In this last example we are essentially saying that the consecutive difference in the 'scale two' information in the third and last quarters is four.

So far we have avoided using the word *wavelet* in our description of the multiscale algorithm above. However, the $d_{j,k}$ 'detail' coefficients are *wavelet* coefficients and the $c_{j,k}$ coefficients are known as *father wavelet* or *scaling function* coefficients. The algorithm that we have derived is one kind of (discrete) wavelet transform (DWT), and the general pyramid algorithm for wavelets is due to Mallat (1989b). The wavelets underlying the transform above are called *Haar wavelets* after Haar (1910). Welcome to Wavelets!

Inverse. The original sequence can be exactly reconstructed by using only the wavelet coefficients $d_{j,k}$ and the last c_{00}. For example, the inverse formulae to the simple ones in (2.3) and (2.5) are

$$c_{j-1,2k} = (c_{j-2,k} + d_{j-2,k})/2 \qquad (2.6)$$

and

$$c_{j-1,2k-1} = (c_{j-2,k} - d_{j-2,k})/2. \qquad (2.7)$$

Section 2.7.4 gives a full description of the inverse discrete wavelet transform.

Sparsity. A key property of wavelet coefficient sequences is that they are often sparse. For example, suppose we started with the input sequence $(1, 1, 1, 1, 2, 2, 2, 2)$. If we processed this sequence with the algorithm depicted by Figure 2.2, then *all* of the wavelet coefficients at scales one and two would be *exactly* zero. The only non-zero coefficient would be $d_0 = -4$. Hence, the wavelet coefficients are an extremely sparse set. This behaviour is characteristic of wavelets: piecewise smooth functions have sparse representations. The vector we chose was actually piecewise constant, an extreme example of piecewise smooth. The sparsity is a consequence of the unconditional basis property of wavelets briefly discussed in the previous chapter and also of the vanishing moments property of wavelets to be discussed in Section 2.4.

Energy. In the example above the input sequence was $(1, 1, 7, 9, 2, 8, 8, 6)$. This input sequence can be thought to possess an 'energy' or norm as defined by $||y||^2 = \sum_{i=1}^{8} y_i^2$. (See Section B.1.3 for a definition of norm.) Here, the norm of the input sequence is $1+1+49+4+64+64+36 = 219$. The transform wavelet coefficients are (from finest to coarsest) $(0, 2, 6, -2, 14, 4, 6, 42)$. What is the norm of the wavelet coefficients? It is $0+4+36+4+196+16+36+1764 = 2056$. Hence the norm, or energy, of the output sequence is much larger than that of the input. We would like a transform where the 'output energy' is the same as the input. We address this in the next section.

2.1.2 Discrete Haar wavelets

To address the 'energy' problem at the end of the last example, let us think about how we might change Formulae (2.1) and (2.2) so as to conserve energy.

Suppose we introduce a multiplier α as follows. Thus (2.1) becomes

$$d_k = \alpha(y_{2k} - y_{2k-1}), \tag{2.8}$$

and similarly (2.2) becomes

$$c_k = \alpha(y_{2k} + y_{2k-1}). \tag{2.9}$$

Thus, with this mini transform the input (y_{2k}, y_{2k-1}) is transformed into the output (d_k, c_k) and the (squared) norm of the output is

$$\begin{aligned}
d_k^2 + c_k^2 &= \alpha^2(y_{2k}^2 - 2y_{2k}y_{2k-1} + y_{2k-1}^2) + \alpha^2(y_{2k}^2 + 2y_{2k}y_{2k-1} + y_{2k-1}^2) \\
&= 2\alpha^2(y_{2k}^2 + y_{2k-1}^2), \tag{2.10}
\end{aligned}$$

where $y_{2k}^2 + y_{2k-1}^2$ is the (squared) norm of the input coefficients. Hence, if we wish the norm of the output to equal the norm of the input, then we should arrange for $2\alpha^2 = 1$ and hence we should set $\alpha = 2^{-1/2}$. With this normalization the formula for the discrete wavelet coefficients is

$$d_k = (y_{2k} - y_{2k-1})/\sqrt{2}, \tag{2.11}$$

and similarly for the father wavelet coefficient c_k. Mostly we keep this normalization throughout, although it is sometimes convenient to use other normalizations. For example, see the normalization for the Haar–Fisz transform in Section 6.4.6.

We can rewrite (2.11) in the following way:

$$d_k = g_0 y_{2k} + g_1 y_{2k-1}, \tag{2.12}$$

where $g_0 = 2^{-1/2}$ and $g_1 = -2^{-1/2}$, or in the more general form:

$$d_k = \sum_{\ell=-\infty}^{\infty} g_\ell y_{2k-\ell}, \tag{2.13}$$

where

$$g_\ell = \begin{cases} 2^{-1/2} & \text{for } \ell = 0, \\ -2^{-1/2} & \text{for } \ell = 1, \\ 0 & \text{otherwise.} \end{cases} \tag{2.14}$$

Equation (2.13) is similar to a filtering operation with filter coefficients of $\{g_\ell\}_{\ell=-\infty}^{\infty}$.

Example 2.2. If we repeat Example 2.1 with the new normalization, then $d_{2,1} = (y_2 - y_1)/\sqrt{2} = (1 - 1)/\sqrt{2} = 0$, and then for the remaining d coefficients at scale $j = 2$ we obtain $d_{2,2} = (y_4 - y_3)/\sqrt{2} = (9 - 7)/\sqrt{2} = \sqrt{2}$, $d_{2,3} = (y_6 - y_5)/\sqrt{2} = (8 - 2)/\sqrt{2} = 3\sqrt{2}$, and, finally, $d_{2,4} = (y_8 - y_7)/\sqrt{2} = (6 - 8)/\sqrt{2} = -\sqrt{2}$.

Also, $c_{2,1} = (y_2 + y_1)/\sqrt{2} = (1+1)/\sqrt{2} = \sqrt{2}$ and for the others $c_{2,2} = (9+7)/\sqrt{2} = 8\sqrt{2}$, $c_{2,3} = (8+2)/\sqrt{2} = 5\sqrt{2}$, and $c_{2,4} = (6+8)/\sqrt{2} = 7\sqrt{2}$.

The $c_{2,k}$ permit us to find the $d_{1,\ell}$ and $c_{1,\ell}$ as follows: $d_{1,1} = (c_{2,2} - c_{2,1})/\sqrt{2} = (8\sqrt{2} - \sqrt{2})/\sqrt{2} = 7$, $d_{1,2} = (c_{2,4} - c_{2,3})/\sqrt{2} = (7\sqrt{2} - 5\sqrt{2})/\sqrt{2} = 2$, and similarly $c_{1,1} = 9$, $c_{1,2} = 12$.

Finally, $d_{0,1} = (c_{1,2} - c_{1,1})/\sqrt{2} = (12 - 9)/\sqrt{2} = 3\sqrt{2}/2$ and $c_{0,1} = (12 + 9)/\sqrt{2} = 21\sqrt{2}/2$.

Example 2.3. Let us perform the transform described in Example 2.2 in WaveThresh. First, start R and load the WaveThresh library by the command

```
> library("WaveThresh")
```

and now create the vector that contains our input to the transform:

```
> y <- c(1,1,7,9,2,8,8,6)
```

The function to perform the discrete wavelet transform in WaveThresh is called wd. So let us perform that transform and store the answers in an object called ywd:

```
> ywd <- wd(y, filter.number=1, family="DaubExPhase")
```

The wd call here supplies two extra arguments: the filter.number and family arguments that specify the type of wavelet that is used for the transform. Here, the values filter.number=1 and family="DaubExPhase" specify Haar wavelets (we will see why these argument names are used later).

The ywd object returned by the wd call is a *composite* object (or list object). That is, ywd contains many different *components* all giving some useful information about the wavelet transform that was performed. The names of the components can be displayed by using the names command as follows:

```
> names(ywd)
[1] "C"        "D"        "nlevels"  "fl.dbase" "filter"
[6] "type"     "bc"       "date"
```

For example, if one wishes to know what filter produced a particular wavelet decomposition object, then one can type

```
> ywd$filter
```

and see the output

```
$H
[1] 0.7071068 0.7071068

$G
NULL

$name
```

```
[1] "Haar wavelet"

$family
[1] "DaubExPhase"

$filter.number
[1] 1
```

which contains information about the wavelet used for the transform. Another interesting component of the `ywd$filter` object is the H component, which is equal to the vector $(2^{-1/2}, 2^{-1/2})$. This vector is the one involved in the filtering operation, analogous to that in (2.13), that produces the c_k, in other words:

$$c_k = \sum_{\ell=-\infty}^{\infty} h_\ell y_{2k-\ell}, \tag{2.15}$$

where

$$h_\ell = \begin{cases} 2^{-1/2} & \text{for } \ell = 0, \\ 2^{-1/2} & \text{for } \ell = 1, \\ 0 & \text{otherwise.} \end{cases} \tag{2.16}$$

Possibly the most important information contained within the wavelet decomposition object `ywd` are the wavelet coefficients. They are stored in the D component of the object, and they can be accessed directly if desired (see the Help page of `wd` to discover how, and in what order, the coefficients are stored). However, the coefficients are stored in a manner that is efficient for computers, but less convenient for human interpretation. Hence, `WaveThresh` provides a function, called `accessD`, to extract the coefficients from the `ywd` object in a readable form.

Suppose we wished to extract the finest-level coefficients. From Example 2.2 these coefficients are $(d_{2,1}, d_{2,2}, d_{2,3}, d_{2,4}) = (0, \sqrt{2}, 3\sqrt{2}, -\sqrt{2})$. We can obtain the same answer by accessing level two coefficients from the `ywd` object as follows:

```
> accessD(ywd, level=2)
[1]   0.000000 -1.414214 -4.242641   1.414214
```

The answer looks correct except the numbers are the negative of what they should be. Why is this? The answer is that `WaveThresh` uses the filter $g_0 = -2^{-1/2}$ and $g_1 = 2^{-1/2}$ instead of the one shown in (2.13). However, this raises a good point: for this kind of analysis one can use filter coefficients themselves or their negation, and/or one can use the reversed set of filter coefficients. In all these circumstances, one still obtains the same kind of analysis.

Other resolution levels in the wavelet decomposition object can be obtained using the `accessD` function with the `levels` arguments set to one and zero. The $c_{j,k}$ father wavelet coefficients can be extracted using the accessC command, which has an analogous mode of operation.

It is often useful to obtain a picture of the wavelet coefficients. This can be achieved in WaveThresh by merely plotting the coefficients as follows:

```
> plot(ywd)
```

which produces a plot like the one in Figure 2.3.

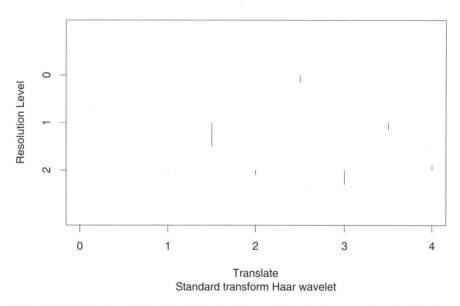

Wavelet Decomposition Coefficients

Fig. 2.3. Wavelet coefficient plot of ywd. The coefficients $d_{j,k}$ are plotted with the finest-scale coefficients at the *bottom* of the plot, and the coarsest at the *top*. The level is indicated by the *left-hand axis*. The value of the coefficient is displayed by a *vertical mark* located along an imaginary horizontal line centred at each level. Thus, the three marks located at resolution level 2 correspond to the three non-zero coefficients $d_{2,2}, d_{2,3}$, and $d_{2,4}$. Note that the zero $d_{2,1}$ is not plotted. The k, or location parameter, of each $d_{j,k}$ wavelet coefficient is labelled 'Translate', and the horizontal positions of the coefficients indicate the approximate position in the original sequence from which the coefficient is derived. Produced by f.wav1().

Other interesting information about the ywd object can be obtained by simply typing the name of the object. For example:

```
> ywd
Class 'wd' : Discrete Wavelet Transform Object:
       ~~ : List with 8 components with names
```

```
        C D nlevels fl.dbase filter type bc date

$C and $D are LONG coefficient vectors

Created on : Mon Dec  4 22:27:11 2006
Type of decomposition:  wavelet

summary(.):
----------

Levels:  3
Length of original:  8
Filter was:  Haar wavelet
Boundary handling:  periodic
Transform type:  wavelet
Date:  Mon Dec  4 22:27:11 2006
```

This output provides a wealth of information the details of which are explained in the WaveThresh Help page for wd.

2.1.3 Matrix representation

The example in the previous sections, and depicted in Figure 2.2, takes a vector input, $y = (1, 1, 7, 9, 2, 8, 8, 6)$, and produces a set of output coefficients that can be represented as a vector:

$$d = (21\sqrt{2}/2, 0, -\sqrt{2}, -3\sqrt{2}, \sqrt{2}, -7, -2, 3\sqrt{2}/2),$$

as calculated at the end of Example 2.2. Since the output has been computed from the input using a series of simple additions, subtractions, and constant scalings, it is no surprise that one can compute the output from the input using a matrix multiplication. Indeed, if one defines the matrix

$$W = \begin{bmatrix} \sqrt{2}/4 & \sqrt{2}/4 & \sqrt{2}/4 & \sqrt{2}/4 & \sqrt{2}/4 & \sqrt{2}/4 & \sqrt{2}/4 & \sqrt{2}/4 \\ 1/\sqrt{2} & -1/\sqrt{2} & 0 & 0 & 0 & 0 & 0 & 0 \\ 0 & 0 & 1/\sqrt{2} & -1/\sqrt{2} & 0 & 0 & 0 & 0 \\ 0 & 0 & 0 & 0 & 1/\sqrt{2} & -1/\sqrt{2} & 0 & 0 \\ 0 & 0 & 0 & 0 & 0 & 0 & 1/\sqrt{2} & -1/\sqrt{2} \\ 1/2 & 1/2 & -1/2 & -1/2 & 0 & 0 & 0 & 0 \\ 0 & 0 & 0 & 0 & 1/2 & 1/2 & -1/2 & -1/2 \\ \sqrt{2}/4 & \sqrt{2}/4 & \sqrt{2}/4 & \sqrt{2}/4 & -\sqrt{2}/4 & -\sqrt{2}/4 & -\sqrt{2}/4 & -\sqrt{2}/4 \end{bmatrix}, \quad (2.17)$$

it is easy to check that $d = Wx$. It is instructive to see the structure of the previous equations contained within the matrix. Another point of interest is in the three 'wavelet vectors' at different scales that are 'stored' within the matrix, for example, $(1/\sqrt{2}, -1/\sqrt{2})$ in rows two through five, $(1/2, 1/2, -1/2, -1/2)$ in rows six and seven, and $(1, 1, 1, 1, -1, -1, -1, -1)/2\sqrt{2}$ in the last row.

The reader can check that W is an orthogonal matrix in that

$$W^T W = W W^T = I. \tag{2.18}$$

One can 'see' this by taking any row and multiplying component-wise by any other row and summing the result (the inner product of any two rows) and obtaining zero for different rows or one for the same row. (See Section B.1.3 for a definition of inner product.)

Since W is an orthogonal matrix it follows that

$$||d||^2 = d^T d = (Wy)^T Wy = y^T (W^T W) y = y^T y = ||y||^2, \tag{2.19}$$

in other words, the length of the output vector d is the same as that of the input vector y and (2.19) is Parseval's relation.

Not all wavelets are orthogonal and there are uses for non-orthogonal wavelets. For example, with non-orthogonal wavelets it is possible to adjust the relative resolution in time and scale (e.g. more time resolution whilst sacrificing frequency resolution), see Shensa (1996) for example. Most of the wavelets we will consider in this book are orthogonal, although sometimes we shall use collections which do not form orthogonal systems, for example, the non-decimated wavelet transform described in Section 2.9.

The operation $d = Wy$ carries out the wavelet transform using a matrix multiplication operation rather than the pyramidal technique we described earlier in Sections 2.1.1 and 2.1.2. If y was a vector containing a dyadic number, $n = 2^J$, of entries and hence W was of dimension $n \times n$, then the computational effort in performing the Wy operation is $\mathcal{O}(n^2)$ (the effort for multiplying the first row of W by y is n multiplications and $n - 1$ additions, roughly n 'operations'. Repeating this for each of the n rows of W results in n^2 operations in total). See Section B.1.9 for a definition of $\mathcal{O}$.

The pyramidal algorithm of earlier sections produces the same wavelet coefficients as the matrix multiplication, but some consideration shows that it produces them in $\mathcal{O}(n)$ operations. Each coefficient is produced with one operation and coefficients are cascaded into each other in an efficient way so that the n coefficients that are produced take only $\mathcal{O}(n)$ operations. This result is quite remarkable and places the pyramid algorithm firmly into the class of 'fast algorithms' and capable of 'real-time' operation. As we will see later, the pyramid algorithm applies to a wide variety of wavelets, and hence one of the advertised benefits of wavelets is that they possess fast wavelet transforms.

The pyramidal wavelet transform is an example of a fast algorithm with calculations carefully organized to obtain efficient operation. It is also the case that only $\mathcal{O}(n)$ memory locations are required for the pyramidal execution as the two inputs can be completely replaced by a father and mother wavelet coefficient at each step, and then the father used in subsequent processing, as in Figure 2.2, for example. Another well-known example of a 'fast algorithm' is the fast Fourier transform (or FFT), which computes the discrete Fourier transform in $\mathcal{O}(n \log n)$ operations. Wavelets have been promoted as being 'faster than the FFT', but one must realize that the discrete wavelet and Fourier transforms compute quite different transforms. Here, $\log n$ is small for even quite large n.

WaveThresh contains functionality to produce the matrix representations of various wavelet transforms. Although the key wavelet transformation functions in WaveThresh, like wd, use pyramidal algorithms for efficiency, it is sometimes useful to be able to obtain a wavelet transform matrix. To produce the matrix W shown in (2.17) use the command GenW as follows:

```
> W1 <-t(GenW(filter.number=1, family="DaubExPhase"))
```

Then examining W1 gives

```
> W1
             [,1]         [,2]         [,3]         [,4]         [,5]
[1,]  0.3535534   0.3535534   0.3535534   0.3535534   0.3535534
[2,]  0.7071068  -0.7071068   0.0000000   0.0000000   0.0000000
[3,]  0.0000000   0.0000000   0.7071068  -0.7071068   0.0000000
[4,]  0.0000000   0.0000000   0.0000000   0.0000000   0.7071068
[5,]  0.0000000   0.0000000   0.0000000   0.0000000   0.0000000
[6,]  0.5000000   0.5000000  -0.5000000  -0.5000000   0.0000000
[7,]  0.0000000   0.0000000   0.0000000   0.0000000   0.5000000
[8,]  0.3535534   0.3535534   0.3535534   0.3535534  -0.3535534
              [,6]         [,7]         [,8]
[1,]   0.3535534   0.3535534   0.3535534
[2,]   0.0000000   0.0000000   0.0000000
[3,]   0.0000000   0.0000000   0.0000000
[4,]  -0.7071068   0.0000000   0.0000000
[5,]   0.0000000   0.7071068  -0.7071068
[6,]   0.0000000   0.0000000   0.0000000
[7,]   0.5000000  -0.5000000  -0.5000000
[8,]  -0.3535534  -0.3535534  -0.3535534
```

which is the same as W given in (2.17) except in a rounded floating-point representation. Matrices for different n can be computed by changing the n argument to GenW and different wavelets can also be specified. See later for details on wavelet specification in WaveThresh.

One can verify the orthogonality of W using WaveThresh. For example:

```
> W1 %*% t(W1)
      [,1] [,2] [,3] [,4] [,5] [,6] [,7] [,8]
[1,]    1    0    0    0    0    0    0    0
[2,]    0    1    0    0    0    0    0    0
[3,]    0    0    1    0    0    0    0    0
[4,]    0    0    0    1    0    0    0    0
[5,]    0    0    0    0    1    0    0    0
[6,]    0    0    0    0    0    1    0    0
```

| [7,] | 0 | 0 | 0 | 0 | 0 | 0 | 1 | 0 |
| [8,] | 0 | 0 | 0 | 0 | 0 | 0 | 0 | 1 |

2.2 Haar Wavelets (on Functions)

2.2.1 Scaling and translation notation

First, we introduce a useful notation. Given any function $p(x)$, on $x \in \mathbb{R}$ say, we can form the (dyadically) scaled and translated version $p_{j,k}(x)$ defined by

$$p_{j,k}(x) = 2^{j/2} p(2^j x - k) \tag{2.20}$$

for all $x \in \mathbb{R}$ and where j, k are integers. Note that if the function $p(x)$ is 'concentrated' around zero, then $p_{j,k}(x)$ is concentrated around $2^{-j}k$. The $2^{j/2}$ factor ensures that $p_{j,k}(x)$ has the same norm as $p(x)$. In other words

$$
\begin{aligned}
||p_{j,k}(x)||^2 &= \int_{-\infty}^{\infty} p_{j,k}^2(x)\, dx \\
&= \int_{-\infty}^{\infty} 2^j p^2(2^j x - k)\, dx \\
&= \int_{-\infty}^{\infty} p^2(y)\, dy = ||p||^2,
\end{aligned}
\tag{2.21}
$$

where the substitution $y = 2^j x - k$ is made at (2.21).

2.2.2 Fine-scale approximations

More mathematical works introduce wavelets that operate on functions rather than discrete sequences. So, let us suppose that we have a function $f(x)$ defined on the interval $x \in [0, 1]$. It is perfectly possible to extend the following ideas to other intervals, the whole line $\mathbb{R}$, or d-dimensional Euclidean space.

Obviously, with a discrete sequence, the finest resolution that one can achieve is that of the sequence itself and, for Haar wavelets, the finest-scale wavelet coefficients involve pairs of these sequence values. For Haar, involving any more than pairs automatically means a larger-scale Haar wavelet. Also, recall that the Haar DWT progresses from finer to coarser scales.

With complete knowledge of a function, $f(x)$, one can, in principle, investigate it at any scale that one desires. So, typically, to initiate the Haar wavelet transform we need to choose a fixed finest scale from which to start. This fixed-scale consideration actually produces a discrete sequence, and further processing of *only the sequence* can produce all subsequent information at coarser scales (although it could, of course, be obtained from the function). We have not answered the question about how to obtain such a discrete sequence from a function. This is an important consideration and there are many ways to do it; see Section 2.7.3 for two suggestions. However, until then suppose that such a sequence, derived from $f(x)$, is available.

In the discrete case the finest-scale wavelet coefficients involved subtracting one element from its neighbour in consecutive pairs of sequence values. For the Haar wavelet transform on functions we derive a similar notion which involves subtracting integrals of the function over consecutive pairs of intervals.

Another way of looking at this is to start with a fine-scale local averaging of the function. First define the *Haar father wavelet* at scale 2^J by $\phi(2^J x)$, where

$$\phi(x) = \begin{cases} 1, & x \in [0, 1], \\ 0 & \text{otherwise.} \end{cases} \tag{2.22}$$

Then define the finest-level (scale 2^J) father wavelet coefficients to be

$$c_{J,k} = \int_0^1 f(x) 2^{J/2} \phi(2^J x - k) \, dx, \tag{2.23}$$

or, using our scaling/translation notation, (2.23) becomes

$$c_{J,k} = \int_0^1 f(x) \phi_{J,k}(x) \, dx = \langle f, \phi_{J,k} \rangle, \tag{2.24}$$

the latter representation using an inner product notation.

At this point, it is worth explaining what the $c_{J,k}$ represent. To do this we should explore what the $\phi_{J,k}(x)$ functions look like. Using (2.20) and (2.22) it can be seen that

$$\phi_{J,k}(x) = \begin{cases} 2^{J/2} & x \in [2^{-J}k, 2^{-J}(k+1)], \\ 0 & \text{otherwise.} \end{cases} \tag{2.25}$$

That is, the function $\phi_{J,k}(x)$ is constant over the interval $I_{J,k} = [2^{-J}k, 2^{-J}(k+1)]$ and zero elsewhere. If the function $f(x)$ is defined on $[0, 1]$, then the range of k where $I_{J,k}$ overlaps $[0, 1]$ is from 0 to $2^J - 1$. Thus, the coefficient $c_{J,k}$ is just the integral of $f(x)$ on the interval $I_{J,k}$ (and proportional to the *local average* of $f(x)$ over the interval $I_{J,k}$).

In fact, the set of coefficients $\{c_{J,k}\}_{k=0}^{2^J-1}$ and the associated Haar father wavelets at that scale define an approximation $f_J(x)$ to $f(x)$ defined by

$$f_J(x) = \sum_{k=0}^{2^J-1} c_{J,k} \phi_{J,k}(x). \tag{2.26}$$

Figure 2.4 illustrates (2.26) for three different values of J. Plot a in Figure 2.4 shows a section of some real inductance plethysmography data collected by the Department of Anaesthesia at the Bristol Royal Infirmary which was first presented and described in Nason (1996). Essentially, this time series reflects changes in voltage, as a patient breathes, taken from a measuring device

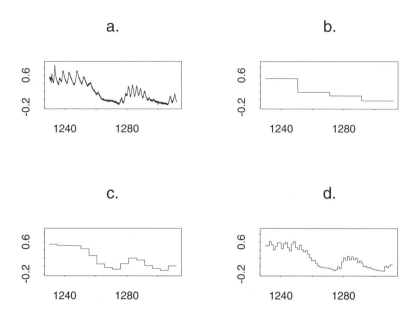

Fig. 2.4. Section of inductance plethysmography data from `WaveThresh` (a), (b) projected onto Haar father wavelet spaces $J = 2$, (c) $J = 4$, and (d) $J = 6$. In each plot the horizontal label is time in seconds, and the vertical axis is milliVolts.

encapsulated in a belt worn by the patient. Plots b, c, and d in Figure 2.4 show Haar father wavelet approximations at levels $J = 2, 4$ and 6. The original data sequence is of length 4096, which corresponds to level $J = 12$. These Haar approximations are reminiscent of the *staircase approximation* useful (for example) in measure theory for proving, among other things, the monotone convergence theorem, see Williams (1991) or Kingman and Taylor (1966).

2.2.3 Computing coarser-scale c from-finer scale ones

Up to now, there is nothing special about J. We could compute the local average over these dyadic intervals $I_{j,k}$ for any j and k. An interesting situation occurs if one considers how to compute the integral of $f(x)$ over $I_{J-1,k}$—that is the interval that is twice the width of $I_{J,k}$ and contains the intervals $I_{J,2k}$ and $I_{J,2k+1}$. It turns out that we can rewrite $c_{J-1,k}$ in terms of $c_{J,2k}$ and $c_{J,2k+1}$ as follows:

$$c_{J-1,k} = \int_{2^{-(J-1)}k}^{2^{-(J-1)}(k+1)} f(x)\phi_{J-1,k}(x)\,dx$$

$$= 2^{-1/2} \int_{2^{-J}2k}^{2^{-J}(2k+2)} f(x)2^{J/2}\phi(2^{J-1}x - k)\,dx \tag{2.27}$$

$$= 2^{-1/2}\left\{ \int_{2^{-J}2k}^{2^{-J}(2k+1)} f(x)2^{J/2}\phi(2^J x - 2k)\,dx \right.$$

$$\left. + \int_{2^{-J}(2k+1)}^{2^{-J}(2k+2)} f(x)2^{J/2}\phi(2^J x - (2k+1))\,dx \right\} \tag{2.28}$$

$$= 2^{-1/2}\left\{ \int_{2^{-J}2k}^{2^{-J}(2k+1)} f(x)\phi_{J,2k}(x)\,dx \right.$$

$$\left. + \int_{2^{-J}(2k+1)}^{2^{-J}(2k+2)} f(x)\phi_{J,2k+1}(x)\,dx \right\}$$

$$= 2^{-1/2}(c_{J,2k} + c_{J,2k+1}). \tag{2.29}$$

The key step in the above argument is the transition from the scale $J - 1$ in (2.27) to scale J in (2.28). This step can happen because, for Haar wavelets,

$$\phi(y) = \phi(2y) + \phi(2y - 1). \tag{2.30}$$

This equation is depicted graphically by Figure 2.5 and shows how $\phi(y)$ is exactly composed of two side-by-side rescalings of the original. Equation (2.30) is a special case of a more general relationship between father wavelets taken at adjacent dyadic scales. The formula for general wavelets is (2.47). It is an important equation and is known as the *dilation equation, two-scale relation*, or the *scaling equation* for father wavelets and it is an example of a *refinement equation*. Using this two-scale relation it is easy to see how (2.27) turns into (2.28) by setting $y = 2^{J-1}x - k$ and then we have

$$\phi(2^{J-1}x - k) = \phi(2^J x - 2k) + \phi(2^J x - 2k - 1). \tag{2.31}$$

A key point here is that to compute $c_{J-1,k}$ one does not necessarily need access to the function and apply the integration given in (2.24). One needs only the values $c_{J,2k}$ and $c_{J,2k+1}$ and to apply the simple Formula (2.29).

Moreover, if one wishes to compute values of $c_{J-2,\ell}$ right down to $c_{0,m}$ (for some ℓ, m), i.e., c at coarser scales still, then one needs only values of c from the next finest scale and the integration in (2.24) is not required. Of course, the computation in (2.29) is *precisely* the one in the discrete wavelet transform that we discussed in Section 2.1.2, and hence computation of *all* the coarser-scale father wavelet coefficients from a given scale 2^J is a fast and efficient algorithm.

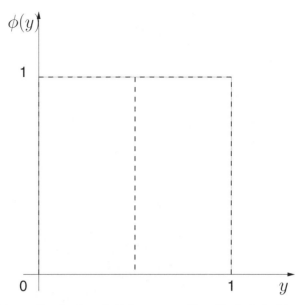

Fig. 2.5. *Solid grey line* is plot of $\phi(y)$ versus y. Two *black dashed lines* are $\phi(2y)$ and $\phi(2y - 1)$ to the left and right respectively.

2.2.4 The difference between scale approximations — wavelets

Suppose we have two Haar approximations of the same function but at two different scale levels. For definiteness suppose we have $f_0(x)$ and $f_1(x)$, the two coarsest approximations (actually approximation is probably not a good term here if the function f is at all wiggly since coarse representations will not resemble the original). The former, $f_0(x)$, is just a constant function $c_{00}\phi(x)$, a multiple of the father wavelet. The approximation $f_1(x)$ is of the form (2.26), which simplifies here to

$$f_1(x) = c_{1,0}\phi_{1,0}(x) + c_{1,1}\phi_{1,1}(x) = c_{1,0}2^{1/2}\phi(2x) + c_{1,1}2^{1/2}\phi(2x - 1). \quad (2.32)$$

What is the difference between $f_0(x)$ and $f_1(x)$? The difference is the 'detail' lost in going from a finer representation, f_1, to a coarser one, f_0. Mathematically:

$$\begin{aligned} f_1(x) - f_0(x) &= c_{0,0}\phi(x) - 2^{1/2}\{c_{10}\phi(2x) + c_{1,1}\phi(2x - 1)\} \\ &= c_{0,0}\{\phi(2x) + \phi(2x - 1)\} \\ &\quad - 2^{1/2}\{c_{1,0}\phi(2x) + c_{1,1}\phi(2x - 1)\}, \end{aligned} \quad (2.33)$$

using (2.30). Hence

$$f_1(x) - f_0(x) = (c_{0,0} - 2^{1/2}c_{1,0})\phi(2x) + (c_{0,0} - 2^{1/2}c_{1,1})\phi(2x - 1), \quad (2.34)$$

and since (2.29) implies $c_{0,0} = (c_{1,0} + c_{1,1})/\sqrt{2}$, we have

$$f_1(x) - f_0(x) = \{(c_{1,1} - c_{1,0})\phi(2x) + (c_{1,0} - c_{1,1})\phi(2x - 1)\} / \sqrt{2}. \quad (2.35)$$

Now suppose we define

$$d_{0,0} = (c_{1,1} - c_{1,0})/\sqrt{2}, \quad (2.36)$$

so that the difference becomes

$$f_1(x) - f_0(x) = d_{0,0} \{\phi(2x) - \phi(2x - 1)\}. \quad (2.37)$$

At this point, it is useful to define the *Haar mother wavelet* defined by

$$\begin{aligned}
\psi(x) &= \phi(2x) - \phi(2x - 1) \\
&= \begin{cases} 1 & \text{if } x \in [0, \frac{1}{2}), \\ -1 & \text{if } x \in [\frac{1}{2}, 1), \\ 0 & \text{otherwise.} \end{cases}
\end{aligned} \quad (2.38)$$

Then the difference between two approximations at scales one and zero is given by substituting $\psi(x)$ into (2.37), to obtain

$$f_1(x) - f_0(x) = d_{0,0}\psi(x). \quad (2.39)$$

Another way of looking at this is to rearrange (2.39) to obtain

$$f_1(x) = c_{0,0}\phi(x) + d_{0,0}\psi(x). \quad (2.40)$$

In other words, the finer approximation at level 1 can be obtained from the coarser approximation at level 0 *plus* the detail encapsulated in $d_{0,0}$. This can be generalized and works at all levels (simply imagine making everything described above operate at a finer scale and stacking those smaller mother and father wavelets next to each other) and one can obtain

$$\begin{aligned}
f_{j+1}(x) &= f_j(x) + \sum_{k=0}^{2^j - 1} d_{j,k}\psi_{j,k}(x) \\
&= \sum_{k=0}^{2^j - 1} c_{j,k}\phi_{j,k}(x) + \sum_{k=0}^{2^j - 1} d_{j,k}\psi_{j,k}(x). \quad (2.41)
\end{aligned}$$

A Haar father wavelet approximation at finer scale $j+1$ can be obtained using the equivalent approximation at scale j plus the details stored in $\{d_{j,k}\}_{k=0}^{2^j - 1}$.

2.2.5 Link between Haar wavelet transform and discrete version

Recall Formulae (2.29) and (2.36)

$$\begin{aligned}
c_{0,0} &= (c_{1,1} + c_{1,0})/\sqrt{2}, \\
d_{0,0} &= (c_{1,1} - c_{1,0})/\sqrt{2}. \quad (2.42)
\end{aligned}$$

These show that, given the finer sequence $(c_{1,0}, c_{1,1})$, it is possible to obtain the coarser-scale mother and father wavelet coefficients without reference to either the actual mother and father wavelet functions themselves (i.e., $\psi(x), \phi(x)$) or the original function $f(x)$. This again generalizes to all scales. Once the finest-scale coefficients $\{c_{J,k}\}_{k=0}^{2^J-1}$ are acquired, all the coarser-scale father and mother wavelet coefficients can be obtained using the discrete wavelet transform described in Section 2.1.2. Precise formulae for obtaining coarser scales from finer, for all scales, are given by (2.91).

2.2.6 The discrete wavelet transform coefficient structure

Given a sequence $y_1, \ldots, y_n$, where $n = 2^J$, the *discrete wavelet transform* produces a vector of coefficients as described above consisting of the last, most coarse, father wavelet coefficient $c_{0,0}$ and the wavelet coefficients $d_{j,k}$ for $j = 0, \ldots, J-1$ and $k = 0, \ldots, 2^j - 1$.

2.2.7 Some discrete Haar wavelet transform examples

We now show two examples of computing and plotting Haar wavelet coefficients. The two functions we choose are the Blocks and Doppler test functions introduced by Donoho and Johnstone (1994b) and further discussed in Section 3.4. These functions can be produced using the **DJ.EX** function in **WaveThresh**. The plots of the Blocks and Doppler functions, and the wavelet coefficients are shown in Figures 2.6 and 2.7. The code that produced Figure 2.7 in **WaveThresh** was as follows:

```
> yy <- DJ.EX()$doppler

> yywd <- wd(yy, filter.number=1, family="DaubExPhase")

> x <- 1:1024

> oldpar <- par(mfrow=c(2,2))

> plot(x, yy, type="l", xlab="x", ylab="Doppler")
> plot(x, yy, type="l", xlab="x", ylab="Doppler")

> plot(yywd, main="")
> plot(yywd,scaling="by.level", main="")

> par(oldpar)
```

The code for Figure 2.6 is similar but Blocks replaces Doppler.

The coefficients plotted in the bottom rows of Figures 2.6 and 2.7 are the same in each picture. The difference is that the coefficients in the bottom

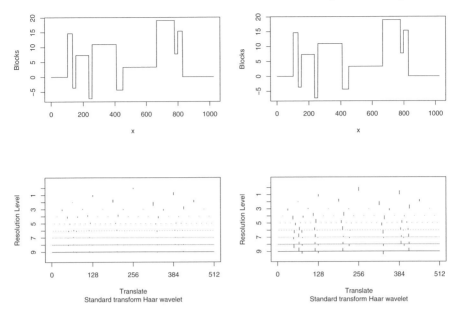

Fig. 2.6. *Top row*: *left* and *right*: identical copies of the Blocks function. *Bottom left*: Haar discrete wavelet coefficients, $d_{j,k}$, of Blocks function (see discussion around Figure 2.3 for description of coefficient layout). All coefficients plotted to same scale and hence different coefficients are comparable. *Bottom right*: as *left* but with coefficients at each level plotted according to a scale that varies according to level. Thus, coefficient size at different levels cannot be compared. The ones at coarse levels are actually bigger. Produced by `f.wav13()`.

left subplot of each are all plotted to the same scale, whereas the ones in the right are plotted with a different scale for each level (by scale here we mean the relative height of the small vertical lines that represent the coefficient values, not the resolution level, j, of the coefficients.) In both pictures it can be seen that as the level increases, to finer scales, the coefficients get progressively smaller (in absolute size). The decay rate of wavelet coefficients is mathematically related to the smoothness of the function under consideration, see Daubechies (1992, Section 2.9), Mallat and Hwang (1992), and Antoniadis and Gijbels (2002).

Three other features can be picked up from these wavelet coefficient plots. In Figure 2.6 the discontinuities in the Blocks function appear clearly as the large coefficients. Where there is a discontinuity a large coefficient appears at a nearby time location, with the exception of the coarser scales where there is not necessarily any coefficient located near to the discontinuities. The other point to note about Figure 2.6 is that many coefficients are *exactly* zero. This is because, in Haar terms, two neighbours, identical in value, were subtracted as in (2.42) to give an exact zero; and this happens at coarser scales too. One

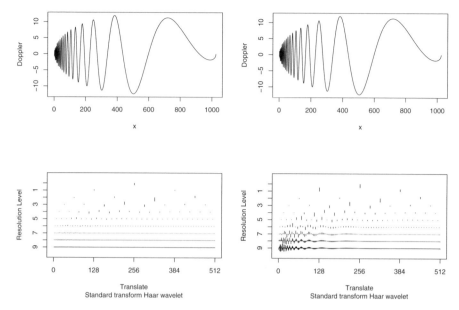

Fig. 2.7. As Figure 2.6 but applied to the Doppler function. Produced by f.wav14().

can examine the coefficients more directly. For example, looking at the first
15 coefficients at level eight gives

```
> accessD(wd(DJ.EX()$blocks), level=8)[1:15]
 [1]   9.471238e-17 -3.005645e-16   1.729031e-15 -1.773625e-16
 [5]   1.149976e-16 -3.110585e-17   4.289763e-18 -1.270489e-19
 [9]  -1.362097e-20  0.000000e+00   0.000000e+00  0.000000e+00
[13]   0.000000e+00  0.000000e+00   0.000000e+00
```

Many of these are exactly zero. The ones that are extremely small (e.g. the
first 9.47×10^{-17}) are only non-zero because the floating-point rounding error
can be considered to be exactly zero for practical purposes. Figure 2.6 is a
direct illustration of the *sparsity* of a wavelet representation of a function
as few of the wavelet coefficients are non-zero. This turns out to happen for
a wide range of signals decomposed with the right kind of wavelets. Such
a property is of great use for compression purposes, see e.g. Taubman and
Marcellin (2001), and for statistical nonparametric regression, which we will
elaborate on in Chapter 3.

Finally, in Figure 2.7, in the bottom right subplot, the oscillatory nature
of the Doppler signal clearly shows up in the coefficients, especially at the
finer scales. In particular, it can be seen that there is a relationship between
the local frequency of oscillation in the Doppler signal and where interesting
behaviour in the wavelet coefficients turns up. Specifically, large variation in
the fine-scale coefficients occurs at the beginning of the set of coefficients.
The 'fine-scale' coefficients correspond to identification of 'high-frequency'

information, and this ties in with the high frequencies in Doppler near the start of the signal. However, large variation in coarser-level coefficients starts much later, which ties in with the lower-frequency part of the Doppler signal. Hence, the coefficients here are a kind of 'time-frequency' display of the varying frequency information contained within the Doppler signal. At a given time-scale location, (j, k), pair, the size of the coefficients gives information on how much oscillatory power there is *locally* at that scale. From such a plot one can clearly appreciate that there is a direct, but reciprocal, relationship between scale and frequency (e.g. small scale is equivalent to high frequency, and *vice versa*). The reader will not then be surprised to learn that these kinds of coefficient plots, and developments thereof, are useful for time series analysis and modelling. We will elaborate on this in Chapter 5.

2.3 Multiresolution Analysis

This section gives a brief and simple account of multiresolution analysis, which is the theoretical framework around which wavelets are built. This section will concentrate on introducing and explaining concepts. We shall quote some results without proof. Full, comprehensive, and mathematical accounts can be found in several texts such as Mallat (1989a,b), Meyer (1993b), and Daubechies (1988, 1992).

The previous sections were prescient in the sense that we began our discussion with a vector of data and, first, produced a set of detail coefficients and a set of smooth coefficients (by differencing and averaging in pairs). It can be appreciated that a function that has reasonable non-zero 'fine-scale' coefficients potentially possesses a more intricate structure than one whose 'fine-scale' coefficients are very small or zero. Further, one could envisage beginning with a low-resolution function and then progressively adding finer detail by inventing a new layer of detail coefficients and working back to the sequence that would have produced them (actually the inverse wavelet transform).

2.3.1 Multiresolution analysis

These kinds of considerations lead us on to 'scale spaces' of functions. Informally, we might define the space V_j as the space (collection) of functions with detail up to some finest scale of resolution. These spaces could possibly contain functions with less detail, but there would be some absolute maximum level of detail. Here larger j would indicate V_j containing functions with finer and finer scales. Hence, one would expect that if a function was in V_j, then it must also be in V_ℓ for $\ell > j$. Mathematically this is expressed as $V_j \subset V_\ell$ for $\ell > j$. This means that the spaces form a ladder:

$$\cdots \subset V_{-2} \subset V_{-1} \subset V_0 \subset V_1 \subset V_2 \subset \cdots . \tag{2.43}$$

As j becomes large and positive we include more and more functions of increasingly finer resolution. Eventually, as j tends to infinity we want to include all functions: mathematically this means that the union of all the V_j spaces is equivalent to the whole function space we are interested in. As j becomes large and negative we include fewer and fewer functions, and detail is progressively lost. As j tends to negative infinity the intersection of all the spaces is just the zero function.

The previous section using Haar wavelets was also instructive as it clearly showed that the detail added at level $j + 1$ is somehow twice as fine as the detail added at level j. Hence, this means that if $f(x)$ is a member of V_j, then $f(2x)$ (which is the same function but varies twice as rapidly as $f(x)$) should belong to V_{j+1}. We refer to this as *interscale linkage*. Also, if we take a function $f(x)$ and shift it along the line, say by an integral amount k, to form $f(x - k)$, then we do not change its level of resolution. Thus, if $f(x)$ is a member of V_0, then so is $f(x - k)$.

Finally, we have not said much about the contents of any of these V_j spaces. Since the Haar father wavelet function $\phi(x)$ seemed to be the key function in the previous sections for building up functions at various levels of detail, we shall say that $\phi(x)$ is an element of V_0 and go further to assume that $\{\phi(x - k)\}_k$ is an orthonormal basis for V_0. Hence, because of interscale linkage we can say that

$$\{\phi_{j,k}(x)\}_{k \in \mathbb{Z}} \text{ forms an orthonormal basis for } V_j. \qquad (2.44)$$

The conditions listed above form the basis for a *multiresolution analysis* (MRA) of a space of functions. The challenge for wavelet design and development is to find such $\phi(x)$ that can satisfy these conditions for a MRA, and sometimes possess other properties, to be useful in various circumstances.

2.3.2 Projection notation

Daubechies (1988) introduced a projection operator P_j that projects a function into the space V_j. Since $\{\phi_{j,k}(x)\}_k$ is a basis for V_j, the projection can be written as

$$f_j(x) = \sum_{k \in \mathbb{Z}} c_{j,k} \phi_{j,k}(x) = P_j f \qquad (2.45)$$

for some coefficients $\{c_{j,k}\}_k$. We saw this representation previously in (2.26) applying to just Haar wavelets. Here, it is valid for more general father wavelet functions, but the result is similar. Informally, $P_j f$ can be thought of as the 'explanation' of the function f using just the father wavelets at level j, or, in slightly more statistical terms, the 'best fitting model' of a linear combination of $\phi_{j,k}(x)$ to $f(x)$ (although this is a serious abuse of terminology because (2.45) is a mathematical representation and not a stochastic one).

The orthogonality of the basis means that the coefficients can be computed by

$$c_{j,k} = \int_{-\infty}^{\infty} f(x)\phi_{j,k}(x)\, dx = <f, \phi_{j,k}>, \tag{2.46}$$

where $<,>$ is the usual inner product operator, see Appendix B.1.3.

2.3.3 The dilation equation and wavelet construction

From the ladder of subspaces in (2.43) space V_0 is a subspace of V_1. Since $\{\phi_{1k}(x)\}$ is a basis for V_1, and $\phi(x) \in V_0$, we must be able to write

$$\phi(x) = \sum_{n \in \mathbb{Z}} h_n \phi_{1n}(x). \tag{2.47}$$

This equation is called the *dilation equation*, and it is the generalization of (2.30). The dilation equation is fundamental in the theory of wavelets as its solution enables one to begin building a *general* MRA, not just for Haar wavelets.

However, for Haar wavelets, if one compares (2.47) and (2.30), one can see that the h_n for Haar must be $h_0 = h_1 = 1/\sqrt{2}$.

The dilation equation controls how the scaling functions relate to each other for two consecutive scales. In (2.30) the father wavelet can be constructed by adding two double-scale versions of itself placed next to each other. The general dilation equation in (2.47) says that $\phi(x)$ is constructed by a linear combination, h_n, of double-scale versions of itself. Daubechies (1992) provides a key result that establishes the existence and construction of the wavelets

Theorem 1 (Daubechies (1992), p.135) *If $\{V_j\}_{j \in \mathbb{Z}}$ with ϕ form a multiresolution analysis of $L^2(\mathbb{R})$, then there exists an associated orthonormal wavelet basis $\{\psi_{j,k}(x) : j, k \in \mathbb{Z}\}$ for $L^2(\mathbb{R})$ such that for $j \in \mathbb{Z}$*

$$P_{j+1}f = P_j f + \sum_k <f, \psi_{j,k}> \psi_{j,k}(x). \tag{2.48}$$

One possibility for the construction of the wavelet $\psi(x)$ is

$$\hat{\psi}(\omega) = e^{i\omega/2}\overline{m_0(\omega/2 + \pi)}\hat{\phi}(\omega/2), \tag{2.49}$$

where $\hat{\psi}$ and $\hat{\phi}$ are the Fourier transforms of ψ and ϕ respectively and where

$$m_0(\omega) = \frac{1}{\sqrt{2}} \sum_n h_n e^{-in\omega}, \tag{2.50}$$

or equivalently

$$\psi(x) = \sum_n (-1)^{n-1}\overline{h_{-n-1}}\phi_{1,n}(x). \tag{2.51}$$

The function $\psi(x)$ is known as the mother wavelet. The coefficient in (2.51) is important as it expresses how the wavelet is to be constructed in terms of the (next) finer-scale father wavelet coefficients. This set of coefficients has its own notation:

$$g_n = (-1)^{n-1} h_{1-n}. \tag{2.52}$$

For Haar wavelets, using the values of h_n from before gives us $g_0 = -1/\sqrt{2}$ and $g_1 = 1/\sqrt{2}$.

Daubechies' Theorem 1 also makes clear that, from (2.48), the difference between two projections $(P_{j+1} - P_j)f$ can be expressed as a linear combination of wavelets. Indeed, the space characterized by the orthonormal basis of wavelets $\{\psi_{j,k}(x)\}_k$ is usually denoted W_j and characterizes the detail lost in going from P_{j+1} to P_j.

The representations given in (2.41) (Haar wavelets) and (2.48) (general wavelets) can be telescoped to give a fine-scale representation of a function:

$$f(x) = \sum_{k \in \mathbb{Z}} c_{j_0,k} \phi_{j_0,k}(x) + \sum_{j=j_0}^{\infty} \sum_{k \in \mathbb{Z}} d_{j,k} \psi_{j,k}(x). \tag{2.53}$$

This useful representation says that a general function $f(x)$ can be represented as a 'smooth' or 'kernel-like' part involving the $\phi_{j_0,k}$ and a set of detail representations $\sum_{k \in \mathbb{Z}} d_{j,k} \psi_{j,k}(x)$ accumulating information at a set of scales j ranging from j_0 to infinity. One can think of the first set of terms of (2.53), $\phi_{j_0,k}$, representing the 'average' or 'overall' level of function and the rest representing the detail. The $\phi(x)$ functions are not unlike many kernel functions often found in statistics—especially in kernel density estimation or kernel regression. However, the father wavelets, $\phi(x)$, tend to be used differently in that for wavelets the 'bandwidth' is 2^{j_0} with j_0 chosen on an integral scale, whereas the usual kernel bandwidth is chosen to be some positive real number. It is possible to mix the ideas of 'wavelet level' and 'kernel bandwidth' and come up with a more general representation, such as (4.16), that combines the strengths of kernels and wavelets, see Hall and Patil (1995), and Hall and Nason (1997). We will discuss this more in Section 4.7

2.4 Vanishing Moments

Wavelets can possess a number of vanishing moments: a function $\psi \in L^2(\mathbb{R})$ is said to have m vanishing moments if it satisfies

$$\int x^\ell \psi(x)\, dx = 0, \tag{2.54}$$

for $\ell = 0, \ldots, m-1$ (under certain technical conditions).

Vanishing moments are important because if a wavelet has m vanishing moments, then all wavelet coefficients of any polynomial of degree m or less will be exactly zero. Thus, if one has a function that is quite smooth and only interrupted by the occasional discontinuity or other singularity, then the wavelet coefficients 'on the smooth parts' will be very small or even zero if the behaviour at that point is polynomial of a certain order or less.

This property has important consequences for data compression. If the object to be compressed is mostly smooth, then the wavelet transform of the object will be sparse in the sense that many wavelet coefficients will be exactly zero (and hence their values do not need to be stored or compressed). The non-zero coefficients are those that encode the discontinuities or non-smooth parts. However, the idea is that for a 'mostly smooth' object there will be few non-zero coefficients to compress further.

Similar remarks apply to many statistical estimation problems. Taking the wavelet transform of an object is often advantageous as it results in a sparse representation of that object. Having only a few non-zero coefficients means that there are few coefficients that actually need to be estimated. In terms of information, it is better to have n pieces of data to estimate a *few* coefficients rather than n pieces of data to estimate n coefficients!

The wvmoments function in WaveThresh calculates the moments of wavelets numerically.

2.5 WaveThresh Wavelets (and What Some Look Like)

2.5.1 Daubechies' compactly supported wavelets

One of the most important achievements in wavelet theory was the construction of orthogonal wavelets that were compactly supported but were smoother than Haar wavelets. Daubechies (1988) constructed such wavelets by an ingenious solution of the dilation equation (2.47) that resulted in a family of orthonormal wavelets (several families actually). Each member of each family is indexed by a number N, which refers to the number of vanishing moments (although in some references N denotes the length of h_n, which is twice the number of vanishing moments). WaveThresh contains two families of Daubechies wavelets which, in the package at least, are called the *least-asymmetric* and *extremal-phase* wavelets respectively. The least-asymmetric wavelets are sometimes known as symmlets. Real-valued compact orthonormal wavelets cannot be symmetric or antisymmetric (unless it is the Haar wavelet, see Daubechies (1992, Theorem 8.1.4)) and the least-asymmetric family is a choice that tries to minimize the degree of asymmetry. A deft discussion of the degree of asymmetry (or, more technically, departure from phase linearity) and the phase properties of wavelet filters can be found in Percival and Walden (2000, pp. 108–116). However, both compactly supported complex-valued and biorthogonal wavelets can be symmetric, see Sections 2.5.2 and 2.6.5.

The key quantity for performing fast wavelet transforms is the sequence of filter coefficients $\{h_n\}$. In WaveThresh, the wd function has access to the filter coefficients of various families through the filter.select function. In WaveThresh, the 'extremal-phase' family has vanishing moments ranging from one (Haar) to ten and the 'least-asymmetric' has them from four to ten. Wavelets in these families possess members with higher numbers of vanishing moments, but they are not stored within WaveThresh.

For example, to see the filter coefficients, $\{h_n\}$, for Haar wavelets, we examine the wavelet with filter.number=1 and family="DaubExPhase" as follows:

```
> filter.select(filter.number=1, family="DaubExPhase")
$H
[1] 0.7071068 0.7071068

$G
NULL

$name
[1] "Haar wavelet"

$family
[1] "DaubExPhase"

$filter.number
[1] 1
```

The actual coefficients are stored in the $H component as an approximation to the vector $(1/\sqrt{2}, 1\sqrt{2})$, as noted before. As another example, we choose the wavelet with filter.number=4 and family="DaubLeAsymm" by:

```
> filter.select(filter.number=4, family="DaubLeAsymm")
$H
[1] -0.07576571 -0.02963553  0.49761867  0.80373875
[5]  0.29785780 -0.09921954 -0.01260397  0.03222310

$G
NULL

$name
[1] "Daub cmpct on least asymm N=4"

$family
[1] "DaubLeAsymm"

$filter.number
[1] 4
```

The length of the vector $H is eight, twice the number of vanishing moments.

It is easy to draw pictures of wavelets within WaveThresh. The following draw.default commands produced the pictures of wavelets and their scaling functions shown in Figure 2.8:

```
> oldpar<-par(mfrow=c(2,1))#To plot one fig above the other

> draw.default(filter.number=4, family="DaubExPhase",
+         enhance=FALSE, main="a.")

>draw.default(filter.number=4, family="DaubExPhase",
+         enhance=FALSE, scaling.function=TRUE, main="b.")

> par(oldpar)
```

The draw.default function is the default method for the generic draw function. The generic function, draw(), can be used directly on objects produced by other functions such as wd so as to produce a picture of the wavelet that resulted in a particular wavelet decomposition. The picture of the $N = 10$ 'least-asymmetric' wavelet shown in Figure 2.9 can be produced with similar commands, but using the arguments filter.number=10 and family="DaubExPhase".

a.

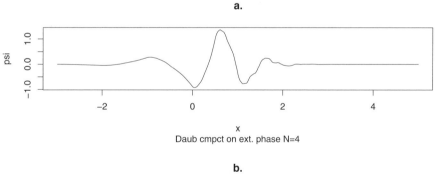

Daub cmpct on ext. phase N=4

b.

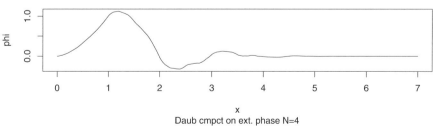

Daub cmpct on ext. phase N=4

Fig. 2.8. Daubechies 'extremal-phase' wavelet with four vanishing moments: (a) mother wavelet and (b) father wavelet. Produced by f.wav2().

a.

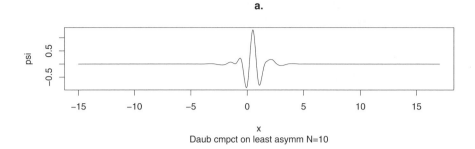

x
Daub cmpct on least asymm N=10

b.

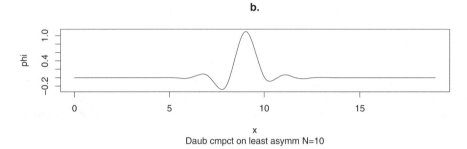

x
Daub cmpct on least asymm N=10

Fig. 2.9. Daubechies 'least-asymmetric' wavelet with ten vanishing moments: (a) mother wavelet, and (b) father wavelet. Produced by `f.wav3()`.

One can also use `GenW` to produce the wavelet transform matrix associated with a Daubechies' wavelet. For example, for the Daubechies' extremal-phase wavelet with two vanishing moments, the associated 8×8 matrix can be produced using the command

```
> W2 <- t(GenW(n=8, filter.number=3, family="DaubExPhase"))
```

and looks like

```
> W2
                [,1]          [,2]          [,3]          [,4]
[1,]   0.35355339    0.35355339    0.35355339    0.35355339
[2,]   0.80689151   -0.33267055    0.00000000    0.00000000
[3,]  -0.13501102   -0.45987750    0.80689151   -0.33267055
[4,]   0.03522629    0.08544127   -0.13501102   -0.45987750
[5,]   0.00000000    0.00000000    0.03522629    0.08544127
[6,]   0.08019599    0.73683030    0.34431765   -0.32938217
[7,]  -0.23056099   -0.04589588   -0.19395265   -0.36155225
[8,]  -0.38061458   -0.02274768    0.21973837    0.55347099
                [,5]          [,6]          [,7]          [,8]
[1,]   0.35355339    0.35355339    0.35355339    0.35355339
[2,]   0.03522629    0.08544127   -0.13501102   -0.45987750
[3,]   0.00000000    0.00000000    0.03522629    0.08544127
```

```
[4,]   0.80689151  -0.33267055   0.00000000   0.00000000
[5,]  -0.13501102  -0.45987750   0.80689151  -0.33267055
[6,]  -0.23056099  -0.04589588  -0.19395265  -0.36155225
[7,]   0.08019599   0.73683030   0.34431765  -0.32938217
[8,]   0.38061458   0.02274768  -0.21973837  -0.55347099
```

2.5.2 Complex-valued Daubechies' wavelets

Complex-valued Daubechies' wavelets (CVDW) are described in detail by Lina and Mayrand (1995). For a given number of N vanishing moments there are 2^{N-1} possible solutions to the equations that define the Daubechies' wavelets, but not all are distinct. When $N = 3$, there are four solutions but only two are distinct: two give the real extremal-phase wavelet and the remaining two are a complex-valued conjugate pair. This $N = 3$ complex-valued wavelet was also derived and illustrated by Lawton (1993) via 'zero-flipping'. Lawton further noted that, apart from the Haar wavelet, the only compactly supported wavelets which are symmetric are CVDWs with an odd number of vanishing moments (other, asymmetric complex-valued wavelets are possible for higher N). The wavelet transform matrix, W, still exists for these complex-valued wavelets but the matrix is now unitary (the complex-valued version of orthogonal), i.e. it satisfies $W\bar{W}^T = \bar{W}^T W = I$, where $\bar{\ }$ denotes complex conjugation.

Currently neither GenW nor draw can produce matrices or pictures of complex-valued wavelets (although it would be not too difficult to modify them to do so). Figure 2.10 shows pictures of the $N = 3$ real- and complex-valued wavelets.

In WaveThresh, the complex-valued wavelet transform is carried out using the usual wd function but specifying the family option to be "LinaMayrand" and using a slightly different specification for the filter.number argument. For example, for these wavelets with five vanishing moments there are four different wavelets which can be used by supplying one of the numbers 5.1, 5.2, 5.3, or 5.4 as the filter.select argument. Many standard WaveThresh functions for processing wavelet coefficients are still available for complex-valued transforms. For example, the plot (or, more precisely, the plot.wd function) function by default plots the modulus of the complex-valued coefficient at each location. Arguments can be specified using the aspect argument to plot the real part, or imaginary part, or argument, or almost any real-valued function of the coefficient.

We show how complex-valued wavelets can be used for denoising purposes, including some WaveThresh examples, in Section 3.14.

2.6 Other Wavelets

There exist many other wavelets and associated multiresolution analyses. Here, we give a quick glimpse of the 'wavelet zoo'! We refer the reader to

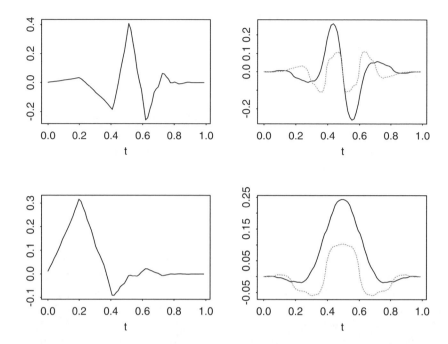

Fig. 2.10. The wavelets (*top*) and scaling functions (*bottom*) for Daubechies' wavelet $N = 3$ (*left*) and complex Daubechies' wavelet equivalent (*right*). The real part is drawn as a *solid black line* and the imaginary part as a *dotted line*.

the comprehensive books by Daubechies (1992) and Chui (1997) for further details on each of the following wavelets.

2.6.1 Shannon wavelet

The Haar scaling function, or father wavelet, given in (2.22) is localized in the x (time or space) domain in that it is compactly supported (i.e., is only non-zero on the interval $[0, 1]$). Its Fourier transform is given by

$$\hat{\phi}(\omega) = (2\pi)^{-1/2} e^{-i\omega/2} \operatorname{sinc}(\omega/2), \qquad (2.55)$$

where

$$\operatorname{sinc}(\omega) = \begin{cases} \frac{\sin \omega}{\omega} & \text{for } \omega \neq 0, \\ 0 & \text{for } \omega = 0. \end{cases} \qquad (2.56)$$

The sinc function is also known as the Shannon sampling function and is much used in signal processing.

Note that $\hat{\phi}(\omega)$ has a decay like $|\omega|^{-1}$. So the Haar mother wavelet is compactly supported in the x domain but with support over the whole of the real line in the frequency domain with a decay of $|\omega|^{-1}$.

For the Shannon wavelet, it is the other way around. The wavelet is compactly supported in the Fourier domain and has a decay like $|x|^{-1}$ in the time domain. Chui (1997, 3.1.5) defines the Shannon father wavelet to be

$$\phi_S(x) = \text{sinc}(\pi x). \tag{2.57}$$

The associated mother wavelet is given by Chui (1997, 4.2.4):

$$\psi_S(x) = \frac{\sin 2\pi x - \cos \pi x}{\pi(x - 1/2)}. \tag{2.58}$$

Both ϕ_S and ψ_S are supported over $\mathbb{R}$. The Fourier transform of ψ_S is given in Chui (1997, 4.2.6) by

$$\hat{\psi}_S(\omega) = -e^{-i\omega/2} I_{[-2\pi, -\pi) \cup (\pi, 2\pi]}(\omega), \tag{2.59}$$

in other words compactly supported on $(\pi, 2\pi]$ and its reflection in the origin.

The Shannon wavelet is not that different from the Littlewood–Paley wavelet given in Daubechies (1992, p. 115) by

$$\psi(x) = (\pi x)^{-1}(\sin 2\pi x - \sin \pi x). \tag{2.60}$$

In statistics the Shannon wavelet seems to be rarely used, certainly in practical applications. In a sense, it is the Fourier equivalent of the Haar wavelet, and hence certain paedagogical statements about wavelets could be made equally about Shannon as about Haar. However, since Haar is easier to convey in the time domain (and possibly because it is older), it is usually Haar that is used. However, the Shannon wavelet is occasionally used in statistics in a theoretical setting. For example, Chui (1997) remarks that Daubechies wavelets, with very high numbers of vanishing moments, 'imitate' the Shannon wavelet, which can be useful in understanding the behaviour of those higher-order wavelets in, for example, estimation of the spectral properties of wavelet-based stochastic processes, see Section 5.3.5.

2.6.2 Meyer wavelet

The Meyer wavelet, see Daubechies (1992, p. 116), has a similar Fourier transform to the Shannon wavelet but with the 'sharp corners' of its compact support purposely smoothed out, which results in a wavelet with faster decay. Meyer wavelets are used extensively in the analysis of statistical inverse problems. Such problems are often expressed as convolution problems which are considerably simplified by application of the Fourier transform, and the compact support of the Meyer wavelet in that domain provides computational benefits. See Kolaczyk (1994, 1996), who first introduced these ideas. For an important recent work that combines fast Fourier and wavelet transforms, and a comprehensive overview of the area see Johnstone et al. (2004). We discuss statistical inverse problems further in Section 4.9.

2.6.3 Spline wavelets

Chui (1997) provides a comprehensive introduction to wavelet theory and to spline wavelets. In particular, Chui (1997) defines the first-order cardinal B-spline by the Haar father wavelet defined in (2.22):

$$N_1(x) = \phi(x). \tag{2.61}$$

The mth-order cardinal B-spline, $m \geq 2$, is defined by the following recursive convolution:

$$N_m(x) = \int_{-\infty}^{\infty} N_{m-1}(x - u)N_1(u)\, du \tag{2.62}$$

$$= \int_0^1 N_{m-1}(x - u)\, du,$$

in view of the definition of N_1.

On taking Fourier transforms since convolutions turn into products, (2.63) turns into:

$$\hat{N}_m(\omega) = \hat{N}_{m-1}(\omega)\hat{N}_1(\omega) = \ldots = \hat{N}_1^m(\omega). \tag{2.63}$$

What is the Fourier transform of $N_1(x)$? We could use (2.55), but it is more useful at this point to take the Fourier transform of both sides of the two-scale Equation (2.30), which in cardinal B-spline notation is

$$N_1(x) = N_1(2x) + N_1(2x - 1), \tag{2.64}$$

and taking Fourier transforms gives

$$\hat{N}_1(\omega) = (2\pi)^{-1/2} \left\{ \int N_1(2x)e^{-i\omega x}\, dx + \int N_1(2x - 1)e^{-i\omega x}\, dx. \right\}$$

$$= \frac{1}{2}(2\pi)^{-1/2} \left\{ \int N_1(y)e^{-iy\omega/2}\, dy + \int N_1(y)e^{-i(y+1)\omega/2}\, dy \right\}$$

$$= \frac{1}{2}(1 + e^{-i\omega/2})\hat{N}_1(\omega/2), \tag{2.65}$$

by substituting $y = 2x$ and $y = 2x - 1$ in the integrals on line 1 of (2.65). Hence using (2.63) and (2.65) together implies that

$$\hat{N}_m(\omega) = \left(\frac{1 + e^{-i\omega/2}}{2} \right)^m \hat{N}_m(\omega/2). \tag{2.66}$$

Chui (1997) shows that (2.66) translates to the following formula in the x domain:

$$N_m(x) = 2^{-m+1} \sum_{k=0}^m \binom{m}{k} N_m(2x - k), \tag{2.67}$$

and this formula defines the two-scale relation for the mth-order cardinal B-spline. For example, for $m = 2$ the two-scale relation (2.67) becomes

$$N_2(x) = 2^{-1} \left\{ N_2(2x) + 2N_2(2x - 1) + N_2(2x - 2) \right\}. \tag{2.68}$$

In view of (2.63) the cardinal B-splines are compactly supported and, using two-scale relations such as (2.67), they can be used as scaling functions to start a multiresolution analysis. The mth-$order$ $cardinal$ $spline$ B-$wavelet$ can be generated by

$$\psi_m(x) = \sum_{k=0}^{3m-2} q_k N_m(2x - k), \tag{2.69}$$

where

$$q_k = \frac{(-1)^k}{2^{m-1}} \sum_{\ell=0}^{m} \binom{m}{\ell} N_{2m}(k - \ell + 1), \tag{2.70}$$

see formulae (5.2.25) and (5.2.24) respectively in Chui (1997). Hence since the cardinal B-splines are compactly supported, the cardinal spline B-wavelet is also compactly supported. However, these spline wavelets are not orthogonal functions, which makes them less attractive for some applications such as nonparametric regression.

The cardinal spline B-wavelets can be orthogonalized according to an 'orthogonalization trick', see Daubechies (1992, p. 147) for details. These orthogonalized wavelets are known as the Battle–Lemarié wavelets. Strömberg wavelets are also a kind of orthogonal spline wavelet with similar properties to Battle–Lemarié wavelets, see Daubechies (1992, p. 116) or Chui (1997, p. 75) for further details.

2.6.4 Coiflets

Coiflets have similar properties to Daubechies wavelets except the scaling function is also chosen so that it has vanishing moments. In other words, the scaling function satisfies (2.54) with ϕ instead of ψ and for moments $\ell = 1, \ldots, m$. Note $\ell = 0$ is not possible since for all scaling functions we must have $\int \phi(x)\,dx \neq 0$. Coiflets are named in honour of R. Coifman, who first requested them, see Daubechies (1992, Section 8.2) for more details.

2.6.5 Biorthogonal wavelets

In what we have seen up to now a wavelet, $\psi(x)$, typically performs both an $analysis$ and a $synthesis$ role. The analysis role means that the wavelet coefficients of a function $f(x)$ can be discovered by

$$d_{j,k} = \int f(x)\psi_{j,k}(x)\,dx. \tag{2.71}$$

Further, the same wavelet can be used to form the *synthesis* of the function as in (2.41). With *biorthogonal wavelets* two functions are used, the analyzing wavelet, $\psi(x)$, and its *dual*, the synthesizing wavelet $\tilde{\psi}(x)$. In regular Euclidean space with an orthogonal basis, one can read off the coefficients of the components of a vector simply by looking at the projection onto the (orthogonal) basis elements. For a non-orthogonal basis, one constructs a dual basis with each dual basis element orthogonal to a corresponding original basis element and the projection onto the dual can 'read off' the coefficients necessary for synthesis. Put mathematically this means that $< \psi_{j,k}, \tilde{\psi}_{\ell,m} >= \delta_{j,\ell}\delta_{k,m}$, see Jawerth and Sweldens (1994).

Filtering systems (filter banks) predating wavelets were known in the signal processing literature, see, e.g., Nguyen and Vaidyanathan (1989), and Vetterli and Herley (1992). For a tutorial introduction to filter banks see Vaidyanathan (1990). The connections to wavelets and development of compactly supported wavelets are described by Cohen et al. (1992).

2.7 The General (Fast) Discrete Wavelet Transform

2.7.1 The forward transform

In Section 2.2.3 we explained how to compute coarser-scale Haar wavelet coefficients. In this section, we will explain how this works for more general wavelet coefficients defined in Section 2.3.

Suppose we have a function $f(x) \in L^2(\mathbb{R})$. How can we obtain coarser-level father wavelet coefficients from finer ones, say, level $J-1$ from J? To see this, recall that the father wavelet coefficients of $f(x)$ at level $J-1$ are given by

$$c_{J-1,k} = \int_{\mathbb{R}} f(x)\phi_{J-1,k}(x)\,dx, \qquad (2.72)$$

since $\{\phi_{J-1,k}(x)\}_k$ is an orthonormal basis for V_{J-1}.

We now need an expression for $\phi_{J-1,k}(x)$ in terms of $\phi_{J,\ell}(x)$ and use the dilation equation (2.47) for this:

$$\begin{aligned}
\phi_{J-1,k}(x) &= 2^{(J-1)/2}\phi(2^{J-1}x - k)\\
&= 2^{(J-1)/2}\sum_n h_n\phi_{1,n}(2^{J-1}x - k)\\
&= 2^{(J-1)/2}\sum_n h_n 2^{1/2}\phi\left\{2(2^{J-1}x - k) - n\right\}\\
&= 2^{J/2}\sum_n h_n\phi(2^J x - 2k - n)\\
&= \sum_n h_n\phi_{J,n+2k}(x). \qquad (2.73)
\end{aligned}$$

In fact, (2.47) is a special case of (2.73) with $J = 1$ and $k = 0$.

Now let us substitute (2.73) into (2.72) to obtain

$$
\begin{aligned}
c_{J-1,k} &= \int_{\mathbb{R}} f(x) \sum_n h_n \phi_{J,n+2k}(x)\, dx \\
&= \sum_n h_n \int_{\mathbb{R}} f(x)\phi_{J,n+2k}(x)\, dx \\
&= \sum_n h_n c_{J,n+2k},
\end{aligned}
\tag{2.74}
$$

or, with a little rearrangement, in its usual form:

$$
c_{J-1,k} = \sum_n h_{n-2k} c_{J,n}.
\tag{2.75}
$$

An equation to obtain *wavelet* coefficients at scale $J-1$ from father wavelet coefficients at scale J can be developed in a similar way. Instead of using the scaling function dilation equation, we use the analogous Equation (2.51) in (2.73), and then after some working we obtain

$$
d_{J-1,k} = \sum_n g_{n-2k} c_{J,n}.
\tag{2.76}
$$

Note that (2.75) and (2.76) hold for any scale j replacing J for $j = 1, \ldots, J$.

2.7.2 Filtering, dyadic decimation, downsampling

The operations described by Equations (2.75) and (2.76) can be thought of in another way. For example, we can achieve the same result as (2.75) by first *filtering* the sequence $\{c_{J,n}\}$ with the filter $\{h_n\}$ to obtain

$$
c^*_{J-1,k} = \sum_n h_{n-k} c_{J,n}.
\tag{2.77}
$$

This is a standard convolution operation. Then we could pick 'every other one' to obtain $c_{J-1,k} = c^*_{J-1,2k}$. This latter operation is known as *dyadic decimation* or *downsampling* by an integer factor of 2. Here, we borrow the notation of Nason and Silverman (1995) and define the (even) dyadic decimation operator $\mathcal{D}_0$ by

$$
(\mathcal{D}_0 x)_\ell = x_{2\ell},
\tag{2.78}
$$

for some sequence $\{x_i\}$.

Hence the operations described by Formulae (2.75) and (2.76) can be written more succinctly as

$$
c_{J-1} = \mathcal{D}_0 \mathcal{H} c_J \text{ and } d_{J-1} = \mathcal{D}_0 \mathcal{G} c_J,
\tag{2.79}
$$

where $\mathcal{H}$ and $\mathcal{G}$ denote the regular filtering operation, e.g. (2.77). In (2.79) we have denoted the input and outputs to these operations using a more efficient vector notation, c_J, c_{J-1}, d_{J-1}, rather than sequences.

Nason and Silverman (1995) note that the *whole* set of discrete wavelet transform (coefficients) can be expressed as

$$d_j = \mathcal{D}_0\mathcal{G}\left(\mathcal{D}_0\mathcal{H}\right)^{J-j-1} c_J, \qquad (2.80)$$

for $j = 0, \ldots, J - 1$ and similarly for the father wavelet coefficients:

$$c_j = \left(\mathcal{D}_0\mathcal{H}\right)^{J-j} c_J, \qquad (2.81)$$

for the same range of j. Remember d_j and c_j here are vectors of length 2^j (for periodized wavelet transforms).

This vector/operator notation is useful, particularly because the computational units $\mathcal{D}_0\mathcal{G}$ and $\mathcal{D}_0\mathcal{H}$ can be compartmentalized in a computer program for easy deployment and robust checking. However, the notation is mathematically liberating and of great use when developing more complex algorithms such as the non-decimated wavelet transform, the wavelet packet transform, or combinations of these. Specifically, one might have wondered why we chose 'even' dyadic decimation, i.e. picked out each even element x_{2j} rather than the odd indexed ones, x_{2j+1}. This is a good question, and the 'solution' *is* the non-decimated transform which we describe in Section 2.9. Wavelet packets we describe in Section 2.11 and non-decimated wavelet packets in Section 2.12.

2.7.3 Obtaining the initial fine-scale father coefficients

In much of the above, and more precisely at the beginning of Section 2.7.1, we mentioned several times that the wavelet transform is initiated from a set of 'finest-scale' father wavelet coefficients, $\{c_{J,k}\}_{k\in\mathbb{Z}}$. Where do these mysterious finest-scale coefficients come from? We outline two approaches.

A *deterministic approach* is described in Daubechies (1992, Chapter 5, Note 12). Suppose the information about our function comes to us as samples, i.e. our information about a function f comes to us in terms of function values at a set of integers: $f(n)$, $n \in \mathbb{Z}$. Suppose that we wish to find the father coefficients of that $f \in V_0$ ('information' orthogonal to V_0 cannot be recovered; whether *your* actual f completely lies in V_0 is another matter).

Now, since $f \in V_0$, we have

$$f(x) = \sum_k < f, \phi_{0,k} > \phi_{0,k}(x), \qquad (2.82)$$

where $< \cdot, \cdot >$ indicates the inner product, again see Appendix B.1.3. Therefore

$$f(n) = \sum_k < f, \phi_{0,k} > \phi(n - k). \qquad (2.83)$$

Applying the discrete Fourier transform (Appendix B.1.7) to both sides of (2.83) gives

$$
\begin{aligned}
\sum_n f(n)e^{-i\omega n} &= \sum_k <f,\phi_{0,k}> \sum_n \phi(n-k)e^{-i\omega n} \\
&= \sum_k <f,\phi_{0,k}> \sum_m \phi(m)e^{-i\omega(m+k)} \\
&= \left\{ \sum_k <f,\phi_{0,k}> e^{-i\omega k} \right\} \left\{ \sum_m \phi(m)e^{-i\omega m} \right\} \\
&= \Phi(\omega) \sum_k <f,\phi_{0,k}> e^{-i\omega k}, \tag{2.84}
\end{aligned}
$$

where $\Phi(\omega) = \sum_m \phi(m)e^{-i\omega m}$ is the discrete Fourier transform of $\{\phi_m(x)\}_m$.

Our objective is to obtain the coefficients $c_{0k} = <f,\phi_{0,k}>$. To do this, rearrange (2.84) and introduce notation $F(\omega)$, to obtain

$$
\sum_k <f,\phi_{0,k}> e^{-i\omega k} = \Phi^{-1}(\omega) \sum_n f(n)e^{-i\omega n} = F(\omega). \tag{2.85}
$$

Hence taking the inverse Fourier transform of (2.85) gives

$$
\begin{aligned}
<f,\phi_{0,k}> &= (2\pi)^{-1} \int_0^{2\pi} F(\omega)e^{i\omega k} d\omega \\
&= (2\pi)^{-1} \int_0^{2\pi} \sum_n f(n)e^{-i\omega(n-k)}\Phi^{-1}(\omega) d\omega \\
&= \sum_n f(n)(2\pi)^{-1} \int_0^{2\pi} e^{-i\omega(n-k)}\Phi^{-1}(\omega) d\omega \\
&= \sum_n a_{n-k}f(n), \tag{2.86}
\end{aligned}
$$

where $a_m = (2\pi)^{-1} \int_0^{2\pi} e^{-i\omega m}\Phi^{-1}(\omega) d\omega$.

For example, for the Daubechies' 'extremal-phase' wavelet with two vanishing moments we have $\phi(0) \approx 0.01$, $\phi(1) \approx 1.36$, $\phi(2) \approx -0.36$, and $\phi(n) = 0, n \neq 0, 1, 2$. This can be checked by drawing a picture of this scaling function. For example, using the WaveThresh function:

```
> draw.default(filter.number=2, family="DaubExPhase",
+     scaling.function=TRUE)
```

Hence denoting $\phi(n)$ by ϕ_n to save space

$$
\Phi(\omega) = \sum_m \phi(m)e^{-i\omega m} \approx \phi_0 + \phi_1 e^{-i\omega} + \phi_2 e^{-2i\omega}, \tag{2.87}
$$

and

$$|\Phi(\omega)|^2 = \phi_0^2 + \phi_1^2 + \phi_2^2 + 2(\phi_0\phi_1 + \phi_1\phi_2)\cos\omega + 2\phi_0\phi_2\cos(2\omega)$$
$$= \phi_1^2 + 2\phi_1\phi_2\cos\omega, \tag{2.88}$$

which is *very* approximately a constant. Here, $a_m = \text{const} \times \delta_{0,m}$ for some constant and $< f, \phi_{0,k} > \approx \text{const} \times f(k)$. So, one might *claim* that one only needs to initialize the wavelet transform using the original function samples. However, it can be seen that the above results in a massive approximation, which is prone to error. Taking the V_0 scaling function coefficients to be the samples is known as the 'wavelet crime', as coined by Strang and Nguyen (1996). The crime can properly be avoided by computing $\Phi(\omega)$ and using more accurate a_m.

A stochastic approach. A somewhat more familiar approach can be adopted in statistical situations. For example, in density estimation, one might be interested in collecting independent observations, $X_1, \ldots, X_n$, from some, unknown, probability density $f(x)$. The scaling function coefficients of f are given by

$$< f, \phi_{j,k} > = \int f(x)\phi_{j,k}(x)\, dx = \mathbb{E}\left[\phi_{j,k}(X)\right]. \tag{2.89}$$

Then an unbiased estimator of $< f, \phi_{j,k} >$ is given by the equivalent sample quantity, i.e.

$$< \widehat{f, \phi_{j,k}} > = n^{-1} \sum_{i=1}^{n} \phi_{j,k}(X_i). \tag{2.90}$$

The values $\phi_{j,k}(X_i)$ can be computed efficiently using the algorithm given in Daubechies and Lagarias (1992). Further details on this algorithm and its use in density estimation can be found in Herrick et al. (2001).

2.7.4 Inverse discrete wavelet transform

In Section 2.2.5, Formula (2.42) showed how to obtain coarser father and mother wavelet coefficients from father coefficients at the next finer scale. These formulae are more usually written for a general scale as something like

$$c_{j-1,k} = (c_{j,2k} + c_{j,2k+1})/\sqrt{2},$$
$$d_{j-1,k} = (c_{j,2k} - c_{j,2k+1})/\sqrt{2}. \tag{2.91}$$

Now suppose our problem is how to invert this operation: i.e. given the $c_{j-1,k}, d_{j-1,k}$, how do we obtain the $c_{j,2k}$ and $c_{j,2k+1}$? One can solve the equations in (2.91) and obtain the following formulae:

$$c_{j,2k} = (c_{j-1,k} + d_{j-1,k})/\sqrt{2},$$
$$c_{j,2k+1} = (c_{j-1,k} - d_{j-1,k})/\sqrt{2}. \tag{2.92}$$

The interesting thing about (2.92) is that the form of the inverse relationship is *exactly* the same as the forward relationship in (2.91).

For general wavelets Mallat (1989b) shows that the inversion relation is given by

$$c_{j,n} = \sum_k h_{n-2k} c_{j-1,k} + \sum_k g_{n-2k} d_{j-1,k}, \qquad (2.93)$$

where h_n, g_n are known as the quadrature mirror filters defined by (2.47) and (2.52). Again, the filters used for computing the inverse transform are the same as those that computed the forward one.

Earlier, in Section 2.1.3, Equation (2.17) displayed the matrix representation of the Haar wavelet transform. We also remarked in that section that the matrix was orthogonal in that $W^T W = I$. This implies that the inverse transform to the Haar wavelet transform is just W^T. For example, the transpose of (2.17) is

$$W^T = \begin{bmatrix} \sqrt{2}/4 & 1/\sqrt{2} & 0 & 0 & 0 & 1/2 & 0 & \sqrt{2}/4 \\ \sqrt{2}/4 & -1/\sqrt{2} & 0 & 0 & 0 & 1/2 & 0 & \sqrt{2}/4 \\ \sqrt{2}/4 & 0 & 1/\sqrt{2} & 0 & 0 & -1/2 & 0 & \sqrt{2}/4 \\ \sqrt{2}/4 & 0 & -1/\sqrt{2} & 0 & 0 & -1/2 & 0 & \sqrt{2}/4 \\ \sqrt{2}/4 & 0 & 0 & 1/\sqrt{2} & 0 & 0 & 1/2 & -\sqrt{2}/4 \\ \sqrt{2}/4 & 0 & 0 & -1/\sqrt{2} & 0 & 0 & 1/2 & -\sqrt{2}/4 \\ \sqrt{2}/4 & 0 & 0 & 0 & 1/\sqrt{2} & 0 & -1/2 & -\sqrt{2}/4 \\ \sqrt{2}/4 & 0 & 0 & 0 & -1/\sqrt{2} & 0 & -1/2 & -\sqrt{2}/4 \end{bmatrix}. \qquad (2.94)$$

Example 2.4. Let us continue Example 2.3, where we computed the discrete Haar wavelet transform on vector y to produce the object ywd. The inverse transform is performed using the wr function as follows:

```
> yinv <- wr(ywd)
```

and if we examine the contents of the inverse transformed vector we obtain

```
> yinv
[1] 1 1 7 9 2 8 8 6
```

So yinv is precisely the same as y, which is exactly what we planned.

2.8 Boundary Conditions

One nice feature of Haar wavelets is that one does not need to think about computing coefficients near 'boundaries'. If one has a dyadic sequence, then the Haar filters transform that sequence in pairs to produce another dyadic sequence, which can then be processed again in the same way. For more general Daubechies wavelets, one has to treat the issue of boundaries more carefully.

For example, let us examine again the simplest compactly supported Daubechies' wavelet (apart from Haar). The detail filter associated with this wavelet has four elements, which we have already denoted in (2.52) by $\{g_k\}_{k=0}^3$. (It is, approximately, $(0.482, -0.837, 0.224, -0.129)$, and can be produced by the `filter.select` function in `WaveThresh`.)

Suppose we have the dyadic data vector $x_0, \ldots, x_{31}$. Then the 'first' coefficient will be $\sum_{k=0}^3 g_k x_k$. Due to even dyadic decimation the next coefficient will be $\sum_{k=0}^3 g_k x_{k+2}$. The operation can be viewed as a window of four g_k consecutive coefficients initially coinciding with the first four elements of $\{x_k\}$ but then skipping two elements 'to the right' each time.

However, one could also wonder what happens when the window also skips to the left, i.e. $\sum_{k=0}^3 g_k x_{k-2}$. Initially, this seems promising as x_0, x_1 are covered when $k = 2, 3$. However, what are x_{-2}, x_{-1} when $k = 0, 1$? Although it probably does not seem to matter very much here as we are only 'missing' two observations (x_{-1}, x_{-2}), the problem becomes more 'serious' for longer filters corresponding to smoother Daubechies' wavelets with a larger number of vanishing moments (for example, with ten vanishing moments the filter is of length 20. So, again we could have x_{-1}, x_{-2} 'missing' but still could potentially make use of the information in $x_0, \ldots, x_{17}$).

An obvious way of coping with this boundary 'problem' is to artificially extend the boundary in some way. In the examples discussed above this consists of artificially providing the 'missing' observations. `WaveThresh` implements two types of boundary extension for some routines: periodic and symmetric end reflection. The function `wd` possesses both options, but many other functions just have the periodic extension. Periodic extension is sometimes also known as being equivalent to using periodized wavelets (for the discrete case).

For a function f defined on the compact interval, say, $[0, 1]$, then periodic extension assumes that $f(-x) = f(1 - x)$. That is information to the 'left' of the domain of definition is actually obtained from the right-hand end of the function. The formula works for both ends of the function, i.e., $f(-0.2) = f(0.8)$ and $f(1.2) = f(0.2)$. Symmetric end reflection assumes $f(-x) = f(x)$ and $f(1 + x) = f(1 - x)$ for $x \in [0, 1]$. To give an example, in the example above x_{-1}, x_{-2} would actually be set to x_{31} and x_{30} respectively for periodic extension and x_1 and x_2 respectively for symmetric end reflection. In `WaveThresh`, these two options are selected using the `bc="periodic"` or `bc="symmetric"` arguments.

In the above we have talked about adapting the data so as to handle boundaries. The other possibility is to leave the data alone and to modify the wavelets themselves. In terms of mathematical wavelets the problem of boundaries occurs when the wavelet, at a coarse scale, is too big, or too big and too near the edge (or over the edge) compared with the interval that the data are defined upon. One solution is to modify the wavelets that overlap the edge by replacing them with special 'edge' wavelets that retain the orthogonality of the system.

The solutions above either wrap the function around on itself (as much as is necessary) for periodized wavelets or reflect the function in its boundaries. The other possibility is to modify the wavelet so that it always remains on the original data domain of definition. This wavelet modification underlies the procedure known as 'wavelets on the interval' due to Cohen et al. (1993). This procedure produces wavelet coefficients at progressively coarser scales but does not borrow information from periodization or reflection. In WaveThresh the 'wavelets on the interval' method is implemented within the basic wavelet transform function, wd, using the bc="interval" option.

2.9 Non-decimated Wavelets

2.9.1 The ϵ-decimated wavelet transform

Section 2.7.2 described the basic forward discrete wavelet transform step as a filtering by $\mathcal{H}$ followed by a dyadic decimation step $\mathcal{D}_0$. Recall that the dyadic decimation step, $\mathcal{D}_0$, essentially picked every even element from a vector. The question was raised there about why, for example, was not every *odd* element picked from the filtered vector instead? The answer is that it could be. For example, we could define the odd dyadic decimation operator $\mathcal{D}_1$ by

$$(\mathcal{D}_1 x)_\ell = x_{2\ell+1}, \tag{2.95}$$

and then the jth level mother and father wavelet coefficients would be obtained by the same formulae as in (2.80) and (2.81), but replacing $\mathcal{D}_0$ by $\mathcal{D}_1$. As Nason and Silverman (1995) point out, this is merely a selection of a different orthogonal basis to the one defined by (2.80) and (2.81).

Nason and Silverman (1995) further point out that, at each level, one could choose either to use $\mathcal{D}_0$ or $\mathcal{D}_1$, and a particular orthogonal basis could be labelled using the zeroes or ones implicit in the choice of particular $\mathcal{D}_0$ or $\mathcal{D}_1$ at each stage. Hence, a particular basis could be represented by the J-digit binary number $\epsilon = \epsilon_{J-1}\epsilon_{J-2}\cdots\epsilon_0$, where ϵ_j is one if $\mathcal{D}_1$ was used to produce level j and zero if $\mathcal{D}_0$ was used. Such a transform is termed the ϵ-*decimated wavelet transform*. Inversion can be handled in a similar way.

Now let us return to the finest scale. It can be easily seen that the effect of $\mathcal{D}_1$ can be achieved by first cyclically 'rotating' the sequence by one position (i.e., making $x_{k+1} = x_k$ and $x_0 = x_{2^J-1}$) and then applying $\mathcal{D}_0$, i.e. $\mathcal{D}_1 = \mathcal{D}_0\mathcal{S}$, where $\mathcal{S}$ is the shift operator defined by $(\mathcal{S}x)_j = x_{j+1}$. By an extension of this argument, and using the fact that $\mathcal{S}\mathcal{D}_0 = \mathcal{D}_0\mathcal{S}^2$, and that $\mathcal{S}$ commutes with $\mathcal{H}$ and $\mathcal{G}$, Nason and Silverman (1995) show that the basis vectors of the ϵ-decimated wavelet transform can be obtained from those of the standard discrete wavelet transform (DWT) by applying a particular shift operator. Hence, they note, the choice of ϵ corresponds to a particular choice of 'origin' with respect to which the basis functions are defined.

An important point is, therefore, that the standard DWT is dependent on choice of origin. A shift of the input data can potentially result in a completely different set of wavelet coefficients compared to those of the original data. For some statistical purposes, e.g., nonparametric regression, we probably would not want our regression method to be sensitive to the choice of origin. Indeed, typically we would prefer our method to be *invariant* to the origin choice, i.e. translation invariant.

2.9.2 The non-decimated wavelet transform (NDWT)

Basic idea. The standard *decimated* DWT is orthogonal and transforms information from one basis to another. The Parseval relation shows that the total energy is conserved after transformation.

However, there are several applications where it might be useful to retain and make use of extra information. For example, in Examples 2.2 on p. 21 coefficient $d_{2,1} = (y_2 - y_1)/\sqrt{2}$ and $d_{2,2} = (y_4 - y_3)/\sqrt{2}$. These first two coefficients encode the difference between (y_1, y_2) and (y_3, y_4) respectively, but what about information that might be contained in the difference between y_2 and y_3? The values y_2, y_3 might have quite different values, and hence not forming a difference between these two values might mean we miss something.

Now suppose we follow the recipe for the ϵ-decimated transform given in the previous section. If the original sequence had been rotated cyclically by one position, then we would obtain the sequence $(y_8, y_1, \ldots, y_7)$, and then on taking the Haar wavelet transform as before gives $d_{2,2} = (y_3 - y_2)/\sqrt{2}$. Applying the transform to the cyclically shifted sequence results in wavelet coefficients, as before, but the set that appeared to be 'missing' as noted above.

Hence, if we wish to retain more information and not 'miss out' potentially interesting differences, we should keep both the original set of wavelet coefficients *and* also the coefficients that resulted after shifting and transformation. However, one can immediately see that keeping extra information destroys the orthogonal structure and the new transformation is redundant. (In particular, one could make use of either the original or the shifted coefficients to reconstruct the original sequence.)

More precisely. The idea of the non-decimated wavelet transform (NDWT) is to retain both the odd and even decimations at each scale and continue to do the same at each subsequent scale. So, start with the input vector $(y_1, \ldots, y_n)$, then apply and retain *both* $\mathcal{D}_0 \mathcal{G} y$ and $\mathcal{D}_1 \mathcal{G} y$—the odd and even indexed 'wavelet' filtered observations. Each of these sequences is of length $n/2$, and so, in total, the number of wavelet coefficients (both decimations) at the finest scale is $2 \times n/2 = n$.

We perform a similar operation to obtain the finest-scale father wavelet coefficients and compute $\mathcal{D}_0 \mathcal{H} y$ ($n/2$ numbers) and $\mathcal{D}_1 \mathcal{H} y$ ($n/2$ numbers). Then for the next level wavelet coefficients we apply *both* $\mathcal{D}_0 \mathcal{G}$ and $\mathcal{D}_1 \mathcal{G}$ to *both* of $\mathcal{D}_0 \mathcal{H} y$ and $\mathcal{D}_1 \mathcal{H} y$. The result of each of these is $n/4$ wavelet coefficients at scale $J - 2$. Since there are four sets, the total number of coefficients is n. A flow diagram illustrating the operation of the NDWT is shown in Figure 2.11.

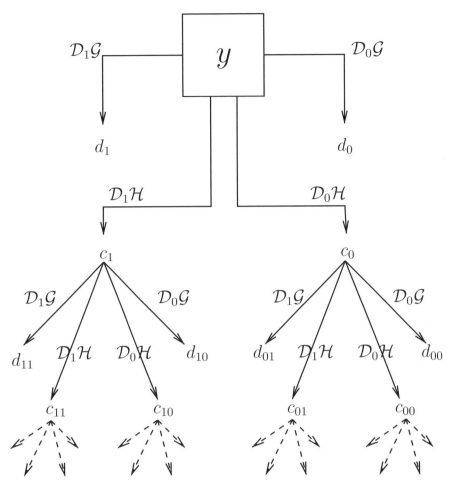

Fig. 2.11. Non-decimated wavelet transform flow diagram. The finest-scale wavelet coefficients are d_0 and d_1. The next finest scale are $d_{00}, d_{01}, d_{10}, d_{11}$. The coefficients that only have 0 in the subscript correspond to the usual wavelet coefficients.

Continuing in this way, at scale $J - j$ there will be 2^j sets of coefficients each of length $2^{-j}n$ for $j = 1, \ldots, J$ (remember $n = 2^J$). For the 'next' coarser scale, there will be twice the number of sets of wavelet coefficients that are half the length of the existing ones. Hence, the number of wavelet coefficients at each scale is always $2^{-j}n \times 2^j = n$. Since there are J scales, the total number of coefficients produced by the NDWT is Jn, and since $J = \log_2 n$, the number of coefficients produced is sometimes written as $n \log_2 n$. Since the production of each coefficient requires a fixed number of operations (which depends on the length of the wavelet filter in use), the computational effort required to compute the NDWT is also $\mathcal{O}(n \log_2 n)$. Although not as 'fast' as the discrete wavelet transform, which is $\mathcal{O}(n)$, the non-decimated algorithm is still considered to be a fast algorithm (the $\log_2 n$ is considered almost to be 'constant').

We often refer to these 'sets' of coefficients as *packets*. These packets are different from the *wavelet packets* described in 2.11, although their method of computation is structurally similar.

Getting rid of the 'origin-sensitivity' is a desirable goal, and many authors have introduced the non-decimated 'technique' working from many points of view and on many problems. See, for example, Holschneider et al. (1989), Beylkin et al. (1991), Mallat (1991), and Shensa (1992). Also, Pesquet et al. (1996) list several papers that innovate in this area. One of the earliest statistical mentions of the NDWT is known as the *maximal-overlap* wavelet transform developed by Percival and Guttorp (1994); Percival (1995). In the latter work, the utility of the NDWT is demonstrated when attempting to estimate the variance within a time series at different scales. We discuss this further in Section 5.2.2. Coifman and Donoho (1995) introduced a NDWT that produced coefficients as 'packets'. They considered different ϵ-decimations as 'cycle spins' and then used the results of averaging over several (often all) cycle spins as a means for constructing a translation-invariant (TI) regression method. We describe TI-denoising in more detail in Section 3.12.1. Nason and Silverman (1995) highlight the possibility for using non-decimated wavelets for determining the spectrum of a nonstationary or evolving time series. This latter idea was put on a sound theoretical footing by Nason et al. (2000), who introduced *locally stationary wavelet processes*: a class of nonstationary evolving time series constructed from non-decimated discrete wavelets, see Section 5.3.

Note that Nason and Silverman (1995) called the NDWT the 'stationary' wavelet transform. This turns out not to be a good name because the NDWT is actually useful for studying nonstationary time series, see Section 5.3. However, some older works occasionally refer to the older name.

2.9.3 Time and packet NDWT orderings

We have already informally mentioned two of the usual ways of presenting, or ordering, non-decimated wavelet coefficients. Let us again return to our simple example of $(y_1, y_2, \ldots, y_8)$. We could simply compute the non-decimated coefficients in *time* order (we omit the $\sqrt{2}$ denominator for clarity):

$$(y_2 - y_1), (y_3 - y_2), (y_4 - y_3), (y_5 - y_4), (y_6 - y_5), (y_7 - y_6), (y_8 - y_7), (y_1 - y_8).$$
$$(2.96)$$

Or we could make direct use of the flow diagram depicted in Figure 2.11 to see the results of the non-decimated transform (to the first scale) as two packets: $\mathcal{D}_0 \mathcal{G}$:

$$(y_2 - y_1), (y_4 - y_3), (y_6 - y_5), (y_8 - y_7), \qquad (2.97)$$

or the odd decimation $\mathcal{D}_1 \mathcal{G}$ packet as

$$(y_3 - y_2), (y_5 - y_4), (y_7 - y_6), (y_1 - y_8). \qquad (2.98)$$

The coefficients contained within (2.96) and both (2.97) and (2.98) are exactly the same; it is merely the orderings that are different. One can continue in either fashion for coarser scales, and this results in a time-ordered NDWT or a packet-ordered one. The time-ordered transform can be achieved via a standard filtering (convolution) operation as noticed by Percival (1995), and hence it is easy to make this work for arbitrary n, not just $n = 2^J$. The packet-ordered transform produces packets as specified by the flow diagram in Figure 2.11.

The time-ordered transform is often useful for time series applications precisely because it is useful to have the coefficients in the same time order as the original data, see Section 5.3. The packet-ordered transform is often useful for nonparametric regression applications as each packet of coefficients corresponds to a particular type of basis element and it is convenient to apply modifications to whole packets and to combine packets flexibly to construct estimators, see Section 3.12.1.

Example 2.5. Let us return again to our simple example. Let $(y_1, \ldots, y_n) = (1, 1, 7, 9, 2, 8, 8, 6)$. In WaveThresh the time-ordered wavelet transform is carried out using, again, the function wd but this time using the argument type="station". For example,

```
> ywdS <- wd(y, filter.number=1, family="DaubExPhase",
+        type="station")
```

computes the NDWT using Haar wavelets. Different wavelets can be selected by supplying values to the filter.number and family arguments as described in Section 2.5.1.

Recall that in Example 2.3 we computed the (decimated) discrete wavelet transform of y and deposited it in the ywd object. Recall also that we extracted the finest-scale wavelet coefficients with the command

```
> accessD(ywd, level=2)
[1]  0.000000 -1.414214 -4.242641  1.414214
```

Let us do the same with our non-decimated object stored in ywdS:

```
> accessD(ywdS, level=2)
[1]  0.000000 -4.242641 -1.414214  4.949747 -4.242641
[6]  0.000000  1.414214  3.535534
```

As emphasized above, see how the original decimated wavelet coefficients appear at positions 1, 3, 5, 7 of the non-decimated vector—these correspond to the even dyadic decimation operator $\mathcal{D}_0$. (Positions 1, 3, 5, 7 are actually odd, but in the C programming language—which much of the low level of WaveThresh is written in—the positions are actually 0, 2, 4, 6. C arrays start at 0 and not 1.)

Example 2.6. Now let us apply the packet-ordered transform. This is carried out using the wst function:

```
> ywst <- wst(y, filter.number=1, family="DaubExPhase")
```

Let us look again at the finest-scale coefficients:

```
> accessD(ywst, level=2)
[1]   0.000000 -1.414214 -4.242641   1.414214 -4.242641
[6]   4.949747  0.000000   3.535534
```

Thus, like the previous example, the number of coefficients at the finest scale is eight, the same as the length of y. However, here the first four coefficients are just the even-decimated wavelet coefficients (the same as the decimated wavelet coefficients from ywd) and the second four are the oddly decimated coefficients.

Although we have accessed the finest-scale coefficients using accessD, since the coefficients in ywdS are packet-ordered, it is more useful to be able to extract packets of coefficients. This extraction can be carried out using the getpacket function. For example, to extract the odd-decimated coefficients type:

```
> getpacket(ywst, level=2, index=1)
[1] -4.242641   4.949747   0.000000   3.535534
```

and use index=0 to obtain the even-decimated coefficients.

What about packets at coarser levels? In Figure 2.11, at the second finest scale ($J - 2$, if $J = 3$ this is level 1), there should be four packets of length 2 which are indexed by binary 00, 01, 10, and 11. These can be obtained by supplying the level=1 argument and setting the index argument to be the base ten equivalent of the binary 00, 01, 10, or 11. For example, to obtain the 10 packet type:

```
> getpacket(ywst, level=1, index=3)
[1] -2.5 -0.5
```

Example 2.7. We have shown above that the time-ordered and packet-ordered NDWTs are equivalent; it is just the orderings that are different. Hence, it should be possible to easily convert one type of object into another. This is indeed the case. For example, one could easily obtain the finest-scale time-ordered coefficients merely by interweaving the two sets of packet-ordered coefficients. Similar weavings operate at different scales, and details can be found in Nason and Sapatinas (2002). In WaveThresh, the conversion between one object and another is carried out using the convert function. Used on a wst class object it produces the wd class object and *vice versa.*

For example, if we again look at the finest-scale coefficients of the ywst object *after* conversion to a wd object, then we should observe the same coefficients as if we applied accessD directly to ywd. Thus, to check:

```
> accessD(convert(ywst), level=2)
[1]  0.000000 -4.242641 -1.414214  4.949747 -4.242641
[6]  0.000000  1.414214  3.535534
```

which gives the same result as applying `accessD` to `ywd`, as shown in Examples 2.5.

Example 2.8. Let us end this series of examples with a more substantial one. Define the symmetric chirp function by

$$y(x) = \sin(\pi/x),$$

for $x = \epsilon' + (-1, -1 + \delta, -1 + 2\delta, \ldots, 1 - 2\delta)$, where $\epsilon' = 10^{-5}$ and $\delta = 1/512$ (essentially x is just a vector ranging from -1 to 1 in increments of $1/512$. The ϵ' is added so that x is never zero. The length of x is 1024). A plot of (x, y) is shown in Figure 2.12. The WaveThresh function `simchirp` can be

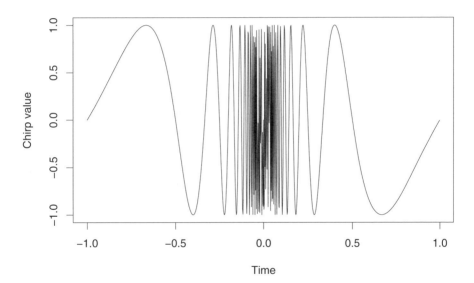

Fig. 2.12. Simulated chirp signal, see text for definition. Produced by `f.wav6()`. (Reproduced with permission from Nason and Silverman (1995).)

used to compute this function and returns an (x, y) vector containing values as follows:

```
> y <- simchirp()

> ywd <- wd(y$y, filter.number=2, family="DaubExPhase")

> plot(ywd, scaling="by.level", main="")
```

These commands also compute the discrete wavelet transform of y using the Daubechies compactly supported extremal-phase wavelet with two vanishing moments and then plot the result which is shown in Figure 2.13. The chirp

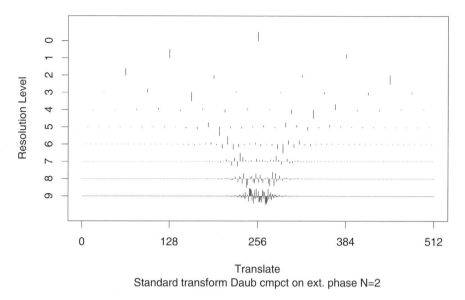

Fig. 2.13. Discrete wavelet coefficients of simulated chirp signal. Produced by f.wav7(). (Reproduced with permission from Nason and Silverman (1995).)

nature of the signal can be clearly identified from the wavelet coefficients, especially at the finer scales. However, as the scales get coarser (small resolution level) it is difficult to see any oscillation, which is unfortunate as the chirp contains power at lower frequencies.

The 'missing' oscillation turns up in its full glory when one examines a non-decimated DWT of the simulated chirp signal. This is shown in Figure 2.14, which was produced using the following code:

```
> ywd <- wd(y$y, filter.number=2, family="DaubExPhase",
+     type="station")

> plot(ywd, scaling="by.level", main="")
```

The reason the lower-frequency oscillation appears to be missing in the DWT is that the transform has been highly decimated at the lower levels (lower frequencies = coarser scales). In comparing Figure 2.13 with 2.14, one can see why the non-decimated transform is more useful for time series analysis. Although the transform is not orthogonal, and the system is redundant, significant information about the oscillatory behaviour at medium and low frequencies (coarser scales) is retained. The chirp signal is an example of a

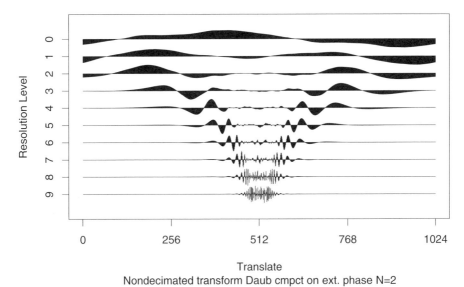

Fig. 2.14. Time-ordered non-decimated wavelet coefficients of simulated chirp signal. Produced by `f.wav8()`. (Reproduced with permission from Nason and Silverman (1995).)

deterministic time series. However, the NDWT is useful in the modelling and analysis of stochastic time series as described further in Chapter 5.

Finally, we also compute and plot the packet-ordered NDWT. This is achieved with the following commands:

```
> ywst <- wst(y$y, filter.number=2, family="DaubExPhase")

> plot(ywst, scaling="by.level", main="")
```

The plot is shown in Figure 2.15. The bottom curve in Figure 2.15 is again just the simulated chirp itself (which can be viewed as finest-scale, data-scale, scaling function coefficients). At the finest detail scale, level nine, there are two packets, the even and oddly decimated coefficients respectively. The packets are separated by a short vertical dotted line. As mentioned above, if one interlaced the coefficients from each packet one at a time, then one would recover the scale level nine coefficients from the time-ordered plot in Figure 2.14. On successively coarser scales the number of packets doubles, but the number of coefficients per packet halves: overall, the number of coefficients remains constant at each level.

2.9.4 Final comments on non-decimated wavelets

To conclude this section on non-decimated wavelets, we refer forward to three sections that take this idea further.

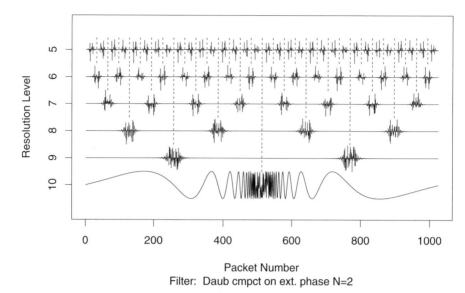

Packet Number
Filter: Daub cmpct on ext. phase N=2

Fig. 2.15. Packet-ordered non-decimated wavelet coefficients of simulated chirp signal. Produced by `f.wav9()`.

1. Section 2.11 describes a generalization of wavelets, called wavelet packets. Wavelet packets can also be extended to produce a non-decimated version, which we describe in Section 2.12.
2. The next chapter explains how the NDWT can be a useful tool for nonparametric regression problems. Section 3.12.1 explains how ϵ-decimated bases can be selected, or how averaging can be carried out over all ϵ-decimated bases in an efficient manner to perform nonparametric regression.
3. Chapter 5 describes how non-decimated wavelets can be used for the modelling and analysis of time series.

Last, we alert the reader to the fact that wavelet transforms computed with different computer packages can sometimes give different results. With decimated transforms the results can be different between packages, although the differences are often minor or trivial and usually due to different wavelet scalings or reflections (e.g., if $\psi(x)$ is a wavelet, then so is $\psi(-x)$). However, with non-decimated transforms the scope for differences increases mainly due to the number of legitimate, but different, ways in which the coefficients can be interwoven.

2.10 Multiple Wavelets

Multiple wavelets are bases with more than one mother and father wavelet. The number of mother wavelets is often denoted by L, and for simplicity of exposition we concentrate on $L = 2$. In this section we base our exposition on, and borrow notation from, Downie and Silverman (1998), which draws on work on multiple wavelets by Geronimo et al. (1994), Strang and Strela (1994), Strang and Strela (1995) and Xia et al. (1996) and Strela et al. (1999). See Goodman and Lee (1994), Chui and Lian (1996), Rong-Qing et al. (1998) for further insights and references.

An (orthonormal) multiple wavelet basis admits the following representation, which is a multiple version of (2.53):

$$f(x) = \sum_{k \in \mathbb{Z}} C_{J,k}^T \Phi_{J,k}(x) + \sum_{j=1}^{J} \sum_{k \in \mathbb{Z}} D_{j,k}^T \Psi_{j,k}(x), \qquad (2.99)$$

where $C_{J,k} = (c_{J,k,1}, c_{J,k,2})^T$ and $D_{j,k} = (d_{j,k,1}, d_{j,k,2})^T$ are *vector* coefficients of dimension $L = 2$. Also, $\Psi_{j,k}(x) = 2^{j/2}\Psi(2^j x - k)$, similarly for $\Phi_{J,k}(x)$, which is very similar to the usual dilation/translation formula, as for single wavelets in (2.20).

The quantity $\Phi(x)$ is actually a vector function of x given by $\Phi(x) = (\phi_1(x), \phi_2(x))^T$ and $\Psi(x) = (\psi_1(x), \psi_2(x))^T$. The basis functions are orthonormal, i.e.

$$\int \psi_l(2^j x - k)\psi_{l'}(2^{j'} x - k') \, dx = \delta_{l,l'}\delta_{j,j'}\delta_{k,k'}, \qquad (2.100)$$

and the $\phi_1(x)$ and $\phi_2(x)$ are orthonormal to all the wavelets $\psi_l(2^j x - k)$. The vector functions $\Phi(x)$ and $\Psi(x)$ satisfy the following dilation equations, which are similar to the single wavelet ones of (2.47) and (2.51):

$$\Phi(x) = \sum_{k \in \mathbb{Z}} H_k \Phi(2x - k), \qquad \Psi(x) = \sum_{k \in \mathbb{Z}} G_k \Phi(2x - k), \qquad (2.101)$$

where now H_k and G_k are 2×2 matrices.

The *discrete multiple wavelet transform* (DMWT), as described by Xia et al. (1996), is similar to the discrete wavelet transform given in (2.75) and (2.76) and can be written as

$$C_{j,k} = \sqrt{2} \sum_n H_n C_{j+1,n+2k} \quad \text{and} \quad D_{j,k} = \sqrt{2} \sum_n G_n C_{j+1,n+2k}, \qquad (2.102)$$

for $j = 0, \ldots, J - 1$. Again, the idea is similar to before: obtain coarser-scale wavelet and scaling function coefficients from finer scale ones. The inverse formula is similar to the single wavelet case.

The rationale for multiple wavelet bases as given by Strang and Strela (1995) is that (i) multiple wavelets can be symmetric, (ii) they can possess short support, (iii) they can have higher accuracy, and (iv) can be orthogonal. Strang and Strela (1995) recall Daubechies (1992) to remind us that no single wavelet can possess these four properties simultaneously.

In most statistical work, the multiple wavelet transform has been proposed for denoising of univariate signals. However, there is immediately a problem with this. The starting (input) coefficients for the DMWT, $\{C_{J,n}\}$, are 2D vectors. Hence, a way has to be found to transform a univariate input sequence into a sequence of 2D vectors. Indeed, such ways have been devised and are called *prefilters*. More on these issues will be discussed in our section on multiple wavelet denoising in Section 3.13.

Example 2.9. Let us continue our previous example and compute the multiple wavelet transform of the chirp signal introduced in Example 2.8. The multiple wavelet code within `WaveThresh` was introduced by Downie (1997). The main functions are: `mwd` for the forward multiple wavelet transform and `mwr` for its inverse. The multiple wavelet transform of the chirp signal can be obtained by the following commands:

```
> y <- simchirp()

> ymwd <- mwd(y$y)

> plot(ymwd, cex=cex)
```

The plot is displayed in Figure 2.16.

2.11 Wavelet Packet Transforms

In Section 2.9 we considered how both odd and even decimation could be applied at each wavelet transform step to obtain the non-decimated wavelet transform. However, for both the decimated and non-decimated transforms the transform cascades by applying filters to the output of a smooth filtering ($\mathcal{H}$). One might reasonably ask the question: is it possible, and sensible, to apply both filtering operations ($\mathcal{H}$ and $\mathcal{G}$) to the output after a filtering by either $\mathcal{H}$ or $\mathcal{G}$? The answer turns out to be yes, and the resulting coefficients are wavelet packet coefficients.

Section 2.3 explained that a set of orthogonal wavelets $\{\psi_{j,k}(x)\}_{j,k}$ was a *basis* for the space of functions $L^2(\mathbb{R})$. However, it is not the only possible basis. Other bases for such function spaces are orthogonal polynomials and the Fourier basis. Indeed, there are many such bases, and it is possible to organize some of them into collections called *basis libraries*. One such library is the *wavelet packet* library, which we will describe below and is described in detail by Wickerhauser (1994), see also Coifman and Wickerhauser (1992) and Hess–Nielsen and Wickerhauser (1996). Other basis libraries include the local cosine basis library, see Bernardini and Kovačević (1996), and the SLEX library which is useful for time series analyses, see Ombao et al. (2001), Ombao et al. (2002, 2005).

Following the description in Coifman and Wickerhauser (1992) we start from a Daubechies mother and father wavelet, ψ and ϕ, respectively. Let $W_0(x) = \phi(x)$ and $W_1(x) = \psi(x)$. Then define the sequence of functions $\{W_k(x)\}_{k=0}^{\infty}$ by

Wavelet Decomposition Coefficients

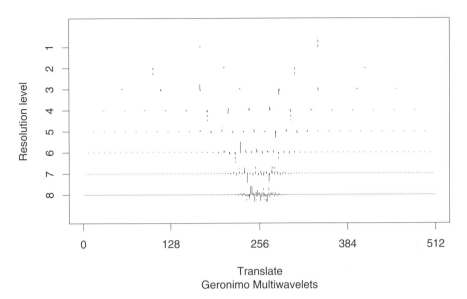

Fig. 2.16. Multiple wavelet transform coefficients of chirp signal. At each time-scale location there are two coefficients: one for each of the wavelets at that location. In WaveThresh on a colour display the two different sets of coefficients can be plotted in different colours. Here, as different line styles, so some coefficients are *dashed*, some are *solid*. Produced by f.wav10().

$$W_{2n}(x) = \sqrt{2} \sum_k h_k W_n(2x - k),$$

$$W_{2n+1}(x) = \sqrt{2} \sum_k g_k W_n(2x - k). \tag{2.103}$$

This definition fulfils the description given above in that both h_k and g_k are applied to $W_0 = \phi$ and both to $W_1 = \psi$ and then both h_k and g_k are applied to the results of these. Coifman and Wickerhauser (1992) define the library of wavelet packet bases to be the collection of orthonormal bases comprised of (dilated and translated versions of W_n) functions of the form $W_n(2^j x - k)$, where $j, k \in \mathbb{Z}$ and $n \in \mathbb{N}$. Here j and k are the scale and translation numbers respectively and n is a new kind of parameter called the number of oscillations. Hence, they conclude that $W_n(2^j - k)$ should be (approximately) centred at $2^j k$, have support size proportional to 2^{-j} and oscillate approximately n times. To form an orthonormal basis they cite the following proposition.

Proposition 1 (Coifman and Wickerhauser (1992)) *Any collection of indices $(j, n, k) \subset \mathbb{N} \times \mathbb{N} \times \mathbb{Z}$, such that the intervals $[2^j n, 2^j (n + 1))$ form a disjoint cover of $[0, \infty)$ and k ranges over all the integers, corresponds to an orthonormal basis of $L^2(\mathbb{R})$.*

In other words, wavelet packets at different scales but identical locations (or covering locations) cannot be part of the same basis.

The definition of wavelet packets in (2.103) shows how coefficients/basis functions are obtained by repeated application of both the $\mathcal{H}$ and $\mathcal{G}$ filters to the original data. This operation is depicted by Figure 2.17. Figure 2.18

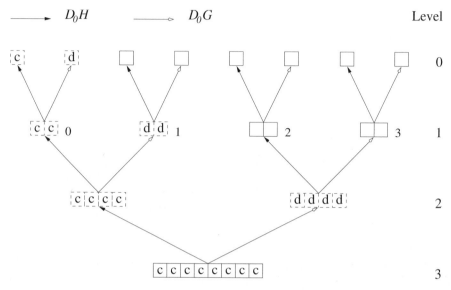

Fig. 2.17. Illustration of wavelet packet transform applied to eight data points (*bottom* to *top*). The $\mathcal{D}_0\mathcal{H}$, $\mathcal{D}_0\mathcal{G}$ filters carry out the smooth and detail operations as in the regular wavelet transform. The difference is that both are applied recursively to the original data with input at the *bottom* of the picture. The regular wavelet coefficients are labelled 'd' and the regular scaling function coefficients are labelled 'c'. The *arrows* at the *top* of the figure indicate which filter is which. Reproduced with permission from Nason and Sapatinas (2002).

shows examples of four wavelet packet functions.

2.11.1 Best-basis algorithms

This section addresses how we might use a library of bases. In Section 2.9.2 we described the set of non-decimated wavelets and how that formed an overdetermined set of functions from which different bases (the ϵ-decimated basis) could be *selected* or, in a regression procedure, representations with respect to many basis elements could be averaged over, see Section 3.12.1. Hence, the non-decimated wavelets are also a basis library and usage usually depends on *selecting* a basis element or *averaging* over the results of many.

For wavelet packets, selection is the predominant mode of operation. Basis averaging could be considered but has received little attention in the

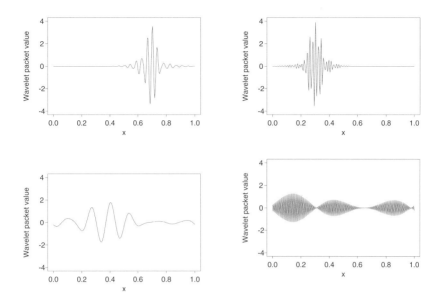

Fig. 2.18. Four wavelet packets derived from Daubechies (1988) least-asymmetric mother wavelet with ten vanishing moments. These four wavelet packets are actually orthogonal and drawn by the `drawwp.default()` function in `WaveThresh`. The vertical scale is exaggerated by ten times. Reproduced with permission from Nason and Sapatinas (2002).

literature. So, for statistical purposes how does selection work? In principle, it is simple for nonparametric regression. One selects a particular wavelet packet basis, obtains a representation of the noisy data with respect to that, thresholds (reduce noise, see Chapter 3), and then inverts the packet transform with respect to that basis. This task can be carried out rapidly using fast algorithms.

However, the whole set of wavelet packet coefficients can be computed rapidly in only $\mathcal{O}(N \log N)$ operations. Hence, an interesting question arises: is it better to select a basis first and then threshold, or is it best to threshold and then select a basis? Again, not much attention has been paid to this problem. For an example of basis selection followed by denoising see Ghugre et al. (2003). However, if the denoising can be done well on all wavelet packet coefficients simultaneously, then it might be better to denoise first and then perform basis selection. The reason for this is that many basis selection techniques are based on the Coifman and Wickerhauser (1992) best-basis algorithm, which is a method that was originally designed to work on deterministic functions. Of course, if the denoising is not good, then the basis selection might not work anyhow. We say a little more on denoising with wavelet packets in Section 3.16.

Coifman–Wickerhauser best-basis method. A possible motivation for the best-basis method is signal compression. That is, can a basis be found that gives the most efficient representation of a signal? Here efficient can roughly be translated into 'most sparse'. A vector of coefficients is said to be sparse if most of its entries are zero, and only a few are non-zero. The Shannon entropy is suggested as a measure of sparsity. Given a set of basis coefficients $\{v_i\}$, the Shannon entropy can be written as $-\sum |v_i|^2 \log |v_i|^2$. For example, the WaveThresh function Shannon.entropy computes the Shannon entropy. Suppose we apply it to two vectors: $v^{(1)} = (0, 0, 1)$ and $v^{(2)} = (1, 1, 1)/\sqrt{3}$. Both these vectors have unit norm.

```
> v1 <- c(0,0,1)
> Shannon.entropy(v1)
[1] 0

> v2 <-  rep(1/sqrt(3), 3)
> Shannon.entropy(v2)
[1] 1.098612
```

(technically Shannon.entropy computes the negative Shannon entropy). These computations suggest that the Shannon entropy is minimized by sparse vectors. Indeed, it can be proved that the 'most-non-sparse' vector $v^{(2)}$ maximizes the Shannon entropy. (Here is a proof for a very simple case. The Shannon entropy is more usually computed on probabilities. Suppose we have two probabilities p_1, p_2 and $p_1 + p_2 = 1$ and the (positive) Shannon entropy is $\mathcal{S}_e(\{p_i\}) = \sum_i p_i \log p_i = p_1 \log p_1 + (1 - p_1) \log(1 - p_1)$. Let us find the stationary points: $\partial \mathcal{S}_e / \partial p_1 = \log p_1 - \log(1 - p_1) = 0$, which implies $\log\{p_1/(1 - p_1)\} = 0$, which implies $p_1 = p_2 = 1/2$, which is the least-sparse vector. Differentiating $\mathcal{S}_e$ again verifies a minimum. For the *negative* Shannon entropy it is a maximum. The proof for general dimensionality $\{p_i\}_{i=1}^n$ is not much more difficult.)

To summarize, the Shannon entropy can be used to measure the sparsity of a vector, and the Coifman–Wickerhauser algorithm searches for the basis that minimizes the overall negative Shannon entropy (actually Coifman and Wickerhauser (1992) is more general than this and admits more general cost functions). Coifman and Wickerhauser (1992) show that the best basis can be obtained by starting from the finest-scale functions and comparing the entropy of that representation by the next coarsest scale packets, and then selecting the one that minimizes the entropy (either the packet or the combination of the two children). Then this operation is applied recursively if required.

2.11.2 WaveThresh example

The wavelet packet transform is implemented in WaveThresh by the wp function. It takes a dyadic-length vector to transform and requires the filter.number and family arguments to specify the underlying wavelet family and number of vanishing moments. For example, suppose we wished to compute the wavelet packet transform of a vector of iid Gaussian random variables. This can be achieved by

```
> z <- rnorm(256)

> zwp <- wp(z, filter.number=2, family="DaubExPhase")

> plot(zwp, color.force=TRUE)
```

This produces the wavelet packet plot shown in Figure 2.19. Let us now replace

Wavelet Packet Decomposition

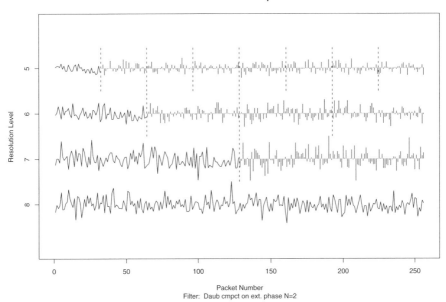

Filter: Daub cmpct on ext. phase N=2

Fig. 2.19. Wavelet packet coefficients of the independent Gaussian sequence z. The time series at the *bottom* of the plot, scale eight, depicts the original data, z. At scales seven through five different wavelet packets are separated by *vertical dotted lines*. The first packet at each scale corresponds to scaling function coefficients, and these have been plotted as a time series rather than a set of small vertical lines (as in previous plots of coefficients). This is because the scaling function coefficients can be thought of as a successive coarsening of the original series and hence are a kind of smooth of the original. The regular wavelet coefficients are always the second packet at each scale. The default plot arguments in `plot.wp` only plot up to scale five and no lower. Produced by `f.wav11()`.

one of the packets in this basis by a very sparse packet. We shall replace the fourth packet (packet 3) at scale six by a packet consisting of all zeroes and a single value of 100. We can investigate the current values of packet $(6,3)$

(index packet 3 is the fourth at scale six, the others are indexed 0, 1, 2) by again using the generic `getpacket` function:

```
> getpacket(zwp, level=6, index=3)
 [1] -1.004520984  2.300091601 -0.765667778  0.614727692
 [5]  2.257342407  0.816656404  0.017121135 -0.353660951
 [9]  0.959106692  1.227197543  ...
 ...
[57]  0.183307351 -0.435437120  0.373848181 -0.565281279
[60] -0.746125550  1.118635271  0.773617722 -1.888108807
[64] -0.182469097
```

So, a vector consisting of a single 100 and all others equal to zero is very sparse. Let us create a new wavelet packet object, `zwp2`, which is identical to `zwp` in all respects except it contains the new sparse packet:

```
> zwp2 <- putpacket(zwp, level=6, index=3,
+     packet=c(rep(0,10), 100, rep(0,53)))
```

```
> plot(zwp2)
```

This last plot command produces the wavelet packet plot as shown in Figure 2.20. To apply the Coifman–Wickerhauser best-basis algorithm using Shannon entropy we use the `MaNoVe` function (which stands for 'make node vector', i.e. select a basis of packet nodes). We can then examine the basis selected merely by typing the name of the node vector:

```
> zwp2.nv <- MaNoVe(zwp2)
```

```
> zwp2.nv
Level:  6  Packet:   3
Level:  3  Packet:   5
Level:  3  Packet:  11
Level:  2  Packet:   5
Level:  2  Packet:  12
 ...
```

As can be seen, $(6,3)$ was selected as a basis element—not surprisingly as it is extremely sparse. The representation can be inverted with respect to the new selected basis contained within `zwp2.nv` by calling `InvBasis(zwp2, zwp2.nv)`. If the inversion is plotted, one sees a very large spike near the beginning of the series. This is the consequence of the 'super-sparse' $(6,3)$ packet.

More information on the usage of wavelet packets in statistical problems in regression and time series can be found in Sections 3.16 and 5.5.

Wavelet Packet Decomposition

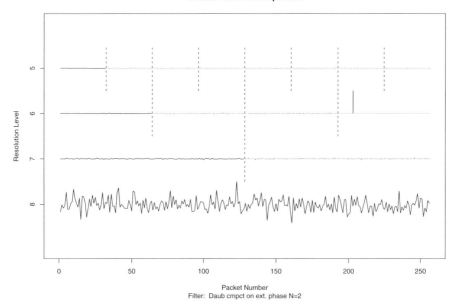

Packet Number

Filter: Daub cmpct on ext. phase N=2

Fig. 2.20. Wavelet packet coefficients of `zwp2`. Apart from packet $(6, 3)$, these coefficients are identical to those in Figure 2.19. However, since the plotted coefficient sizes are relative, the duplicated coefficients have been plotted much smaller than those in Figure 2.19 because of the large relative size of the 100 coefficient in the fourth packet at level 6 (which stands out as the tenth coefficient after the start of the fourth packet indicated by the *vertical dotted line.*) Produced by `f.wav12()`.

2.12 Non-decimated Wavelet Packet Transforms

The discrete wavelet transform relied on even dyadic decimation, $\mathcal{D}_0$, and the smoothing and detail filters, $\mathcal{H}$ and $\mathcal{G}$, but iterating on the results of the $\mathcal{H}$ filter only. One generalization of the wavelet transform, the non-decimated transform, pointed out that the odd dyadic decimation operator, $\mathcal{D}_1$, was perfectly valid and both could be used at each step of the wavelet transform.

In the previous section another generalization, the wavelet packet transform, showed that iteration of both $\mathcal{H}$ and $\mathcal{G}$ could be applied to the results of *both* of the previous filters, not just $\mathcal{H}$.

These two generalizations can themselves be combined by recursively applying the four operators $\mathcal{D}_0\mathcal{H}$, $\mathcal{D}_0\mathcal{G}$, $\mathcal{D}_1\mathcal{H}$, and $\mathcal{D}_1\mathcal{G}$. Although this may sound complicated, the result is that we obtain wavelet packets that are non-decimated. Just as non-decimated wavelets are useful for time series analysis, so are non-decimated wavelet packets. See Section 5.6 for further information.

2.13 Multivariate Wavelet Transforms

The extension of wavelet methods to 2D regularly spaced data (images) and to such data in higher dimensions was proposed by Mallat (1989b). A simplified explanation appears in Nason and Silverman (1994). Suppose one has an $n \times n$ matrix x where n is dyadic. In its simplest form one applies both the $\mathcal{D}_0\mathcal{H}$ and $\mathcal{D}_0\mathcal{G}$ operators from (2.79) to the rows of the matrix. This results in two $n \times (n/2)$ matrices, which we will call H and G. Then both operators are again applied but to both the columns of H and G. This results in four matrices HH, GH, HG, and GG each of dimension $(n/2) \times (n/2)$. The matrix HH is the result of applying the 'averaging' operator $\mathcal{D}_0\mathcal{H}$ to both rows and columns of x, and this is the set of scaling function coefficients with respect to the 2D scaling function $\Phi(x,y) = \Phi(x)\Phi(y)$. The other matrices GH, HG, and GG create finest-scale wavelet detail in the horizontal, vertical, and 'diagonal' directions. This algorithmic step is then repeated by applying the same filtering operations to HH, which generates a new HH, GH, HG, and GG at the next finest scale and then the step is repeated by application to the new HH, and so on (exactly the same as the recursive application of $\mathcal{D}_0\mathcal{H}$ to the c vectors in the 1D transform). The basic algorithmic step for the 2D separable transform is depicted in Figure 2.21.

The transform we have described here is an example of a separable wavelet transform because the 2D scaling function $\Phi(x,y)$ can be separated into the product of two 1D scaling functions $\phi(x)\phi(y)$. The same happens with the wavelets except there are three of them encoding the horizontal, vertical, and diagonal detail $\Psi^H(x,y) = \psi(x)\phi(y)$, $\Psi^V(x,y) = \phi(x)\psi(y)$, and $\Psi^D(x,y) = \psi(x)\psi(y)$. For a more detailed description see Mallat (1998). For nonseparable wavelets see Kovačević and Vetterli (1992) or Li (2005) for a more recent construction and further references.

The 2D transform of an image is shown in Figure 2.22, and the layout of the coefficients is shown in Figure 2.23. The coefficient image was produced with the following commands in WaveThresh:

```
#
# Enable access to teddy image
#
> data(teddy)
#
# Setup grey scale for image colors
#
> greycol <- grey((0:255)/255)
#
# Compute wavelet coefficients of teddy image
#
> teddyimwd <- imwd(teddy, filter.number=10)
#
# Compute scaling for coefficient display
```

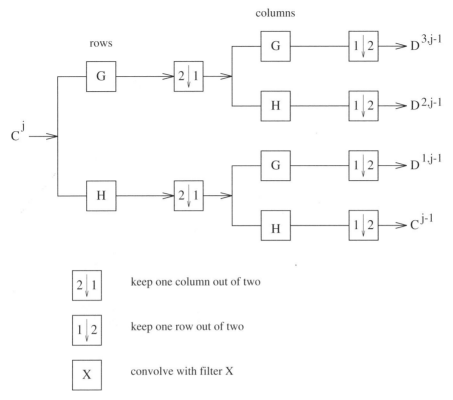

Fig. 2.21. Schematic diagram of the central step of the 2D discrete wavelet transform. The input image on the *left* is at level j and the outputs are the smoothed image C^{j-1} plus horizontal, vertical, and diagonal detail D^1, D^2, and D^3. The smoothed image C^{j-1} is fed into an identical step at the next coarsest resolution level. Here $2 \downarrow 1$ and $1 \downarrow 2$ denote dyadic decimation $\mathcal{D}_0$ in the horizontal and vertical directions. (After Mallat (1989b)).

```
# (just a suggestion)
#
> myt <- function(x) 20+sqrt(x)
#
# Display image of Teddy
#
> plot(teddyimwd, col=greycol, transform=TRUE, tfunction=myt)
```

In both Figures 2.22 and 2.23, the top left block corresponds to the finest detail in the vertical direction, the top right block corresponds to the finest detail in the diagonal direction, and the bottom right block to the horizontal detail (i.e. the GH, GG, and HG coefficients produced by the algorithm mentioned above). These three blocks form an inverted 'L' of coefficients at the finest resolution. The next step of the algorithm produces a similar set of coefficients at the second finest resolution according to the same layout, and so on.

Fig. 2.22. Wavelet coefficients of teddy image shown in Figure 4.6 on p. 142. Produced by f.wav15(). (After Mallat (1989b)).

Within WaveThresh, the 1D DWT is carried out using the wd and wr functions, the 2D DWT using the imwd and imwr functions, and the 3D DWT using the wd3D and wr3D functions. Both wd and imwd can perform the time-ordered non-decimated transform, and wst and wst2D can perform 2D packet-ordered non-decimated transforms.

2.14 Other Topics

The continuous wavelet transform. We have first presented the story of wavelets as a method to extract multiscale information from a sequence, then explained how a set of functions called Haar wavelets can be used to provide a theoretical underpinning to this extraction. Then we demonstrated that the idea could be generalized to wavelets that are smoother than Haar wavelets and, for some applications, more useful. In many mathematical presentations, e.g. Daubechies (1992) (whose development we will follow here), the starting point is the *continuous wavelet transform*, CWT. Here the starting point is a function $f \in L^2(\mathbb{R})$ whose CWT is given by

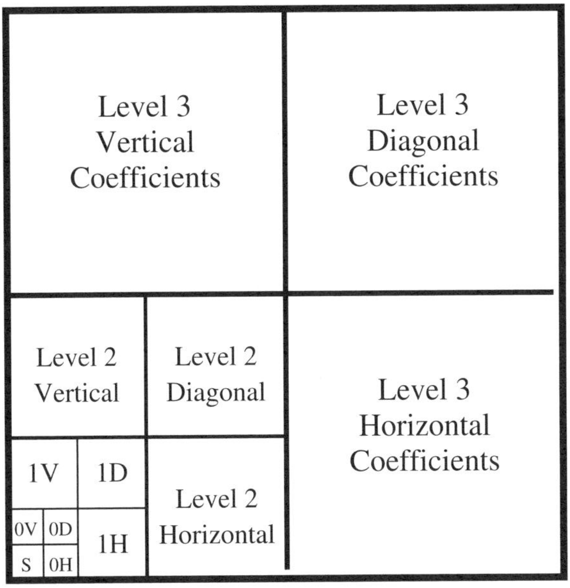

Fig. 2.23. Diagram showing general layout of wavelet coefficients as depicted in Figure 2.22. The plan here stops at the fourth iteration (level 0) whereas the one in Figure 2.22 is the result of nine iterations. (After Mallat (1989b)).

$$F(a,b) = \int_{-\infty}^{\infty} f(x)\psi^{a,b}(x)\,dx, \tag{2.104}$$

for $a, b \in \mathbb{R}, a \neq 0$, where

$$\psi^{a,b}(x) = |a|^{-1/2}\psi\left(\frac{x-b}{a}\right), \tag{2.105}$$

where $\psi \in L^2(\mathbb{R})$ satisfies a technical admissibility condition which Daubechies (1992) notes "for all practical purposes, [the admissibility condition] is equivalent to the requirement that $\int \psi(x)\,dx = 0$". The function f can be recovered from its CWT, $F(a,b)$. There are many accounts of the CWT; see, for example, Heil and Walnut (1989), Daubechies (1992), Meyer (1993b), Jawerth and Sweldens (1994), Vidakovic (1999a), and Mallat (1998), to name but a few. As for the DWT above there are many wavelets that can be used for $\psi(x)$ here, for example, the Haar, the Shannon from Section 2.6.1, and the so-called 'Mexican-hat' wavelet which is the second derivative of the normal probability density function.

Antoniadis and Gijbels (2002) note that in practical applications the CWT is usually computed on a discrete grid of points, and one of the most popular, but by no means the only, discretizations is to set $a = 2^j$ and $b = k$. Antoniadis and Gijbels (2002) refer to this as the continuous discrete wavelet transform (CDWT) and mention a fast computational algorithm by Abry (1994) which is equivalent to the non-decimated wavelet transform from Section 2.9. The CWT can be discretized to $a = 2^j$ and $b = k2^j$ to obtain the DWT. In this context the CDWT is often used for jump or singularity detection, such as in Mallat and Hwang (1992) and Antoniadis and Gijbels (2002) and references therein. Torrence and Compo (1998) is an extremely well-written and engaging description of the use of the CWT for the analysis of meteorological time series such as the El Niño Southern Oscillation.

Lifting is a technique that permits the multiscale method to be applied to more general data situations. The wavelet transforms we have described above are limited to data that occur regularly spaced on a grid, and for computational convenience and speed we have also assumed that $n = 2^J$ (although this latter restriction can often be circumvented by clever algorithm modification). What about data that are not regularly spaced? Most regression methods, parametric and nonparametric, can be directly applied to irregularly spaced data, so what about wavelet methods? Many papers have been written that are devoted to enabling wavelets in the irregular case. This body of work is reviewed in Section 4.5 along with an example of use for one of them. Generally, many of them work by 'transforming' the irregular data, in some way, so as to fit the regular wavelet transform. Lifting is somewhat different as it can cope directly with the irregular data.

As with wavelets we introduce lifting by reexamining the Haar wavelet transform but presented 'lifting style'. Suppose we begin with two data points (or scaling function coefficients at some scale), c_1 and c_2. The usual way of presenting the Haar transform is with equations such as (2.42). However, we could achieve the same result by first carrying the following operation:

$$c_1 \leftarrow (c_1 - c_2)/\sqrt{2}. \tag{2.106}$$

Here $\leftarrow$ is used instead of $=$ to denote that the result of the calculation on the right-hand side of the equation is assigned and overwrites the existing location c_1. Then taking this *new* value of c_1 we can form

$$c_2 \leftarrow \sqrt{2}c_2 + c_1. \tag{2.107}$$

In lifting, Equation (2.106) is known as the *predict* step and (2.107) is known as the *update* step. The steps can be chained similarly to the Haar transform to produce the full transform. The beauty of lifting is its simplicity. For example, the inverse transformation merely reverses the steps by undoing (2.107) and then undoing (2.106). Many other existing wavelet transforms can be 'put in lifting form'. The lifting scheme was introduced by Wim Sweldens, see Sweldens (1996, 1997) for example.

A major benefit of lifting is that the idea can be extended to a wider range of data set-ups than described earlier in this book. For example, for irregular data, in several dimensions, it is possible to obtain the detail, or 'wavelet' coefficient, for a point in the following way. First, identify the neighbours of such a point and then, using some method, e.g., linear regression on the neighbours, work out the fitted value of the point. Then the detail is just the fitted value minus the observed value. These multiresolution analyses for irregular data can be used for nonparametric regression purposes but are beyond the scope of the present text. See Jansen et al. (2001), Claypoole et al. (2003), Delouille et al. (2004a,b), Nunes et al. (2006), and the book by Jansen and Oonincx (2005) for further information on lifting in statistics.

3

Wavelet Shrinkage

3.1 Introduction

Regression is probably the area of statistics that has received the most attention from researchers in statistical wavelet methods. Here, we use the term 'regression' to cover a wide range of tools, theory, and tricks that apply to many kinds of data sets and structures. Wavelet methods are usually employed as a form of nonparametric regression, and the techniques take on many names such as wavelet shrinkage, curve estimation, or wavelet regression. Nonparametric regression itself forms a significant, vibrant area of modern statistics: for example see the following *books*: Härdle (1992), Green and Silverman (1993), Wand and Jones (1994), Fan (1996), Bowman and Azzalini (1997), Simonoff (1998), Eubank (1999), and Wasserman (2005), to name but a few. Antoniadis (2007) provides a recent review of wavelet methods in nonparametric curve estimation and covers the connection between wavelet shrinkage and nonlinear diffusions, penalized least-squares wavelet estimation (not covered in this book), and block thresholding (Section 3.15).

This chapter concentrates on wavelet shrinkage. That is, one observes a function contaminated with additive noise, takes a wavelet transform, modifies, or shrinks the noisy function's wavelet coefficients, and then takes the inverse wavelet transform to estimate the function. The basic, and very popular, model setup is as follows. The general idea is that one obtains observations, $y = (y_1, \ldots, y_n)$, that arise from the following model:

$$y_i = g(x_i) + e_i, \quad \text{for } i = 1, \ldots, n, \tag{3.1}$$

where $x_i = i/n$. The aim is to estimate the unknown function $g(x)$, for $x \in [0, 1]$, using the noisy observations y_i. In the basic model it is usually assumed that $e_i \sim N(0, \sigma^2)$ independently, or white noise (that is independent, with zero mean with constant variance, σ^2). This model contains several assumptions which can be relaxed or extended, see Chapter 4. A multiscale method to deal with multiplicative noise is described in Chapter 6.

Most of the methods described in the following sections actually concentrate on estimating g at the prescribed set of locations x_i, $i = 1, \ldots, n$, although there are ways of obtaining an estimate of $g(x)$ for arbitrary $x \in [0, 1]$. One possible reason why multiscale researchers seldom worry too much about estimation for arbitrary x is that often the sample size, n, is neccesarily large to capture detailed information about local phenomena, and hence any arbitrary practical value of x is close to one of the sample points, x_i. The other point to mention is that we have chosen the domain of g to be $[0, 1]$, but the methods can usually be extended to work on any interval and, indeed, in principle at least, the whole real line.

3.2 Wavelet Shrinkage

The seminal papers by Donoho (1993b, 1995a) Donoho and Johnstone (1994b, 1995), and Donoho et al. (1995) introduced the concept of wavelet shrinkage to the statistical literature. Their general idea is that the discrete wavelet transform (DWT) as described in the previous chapter is applied to model (3.1). Let W denote the particular DWT that we choose (we use the matrix notation here, but the actual calculations are typically performed using Mallat's fast pyramid algorithm) and let y, g, and e represent the vectors of observations, true unknown function, and noise respectively.

Since the DWT is linear, we can write the wavelet-transformed model as

$$d^* = d + \epsilon, \tag{3.2}$$

where $d^* = Wy$, $d = Wg$, and $\epsilon = We$, and where the wavelet matrix W was described in Sections 2.1.3 and 2.5.1.

Three features of (3.2) are central to the success of wavelet shrinkage:

1. For many functions, e.g. smooth functions, smooth functions with some jump discontinuities or other inhomogeneities, the wavelet coefficient vector d is, by the discussion in Chapter 2, a *sparse* vector.
2. Further, because of Parseval's relation, demonstrated by (2.19), the energy in the function domain $\sum_i g(x_i)^2$ is equal to the sum of squares of wavelet coefficients $\sum_{j,k} d_{j,k}^2$. However, taken with the sparsity this means that the 'energy' of the original signal, g, is now concentrated into fewer coefficients and nothing is lost. Hence, relative to the noise variance not only will the vector d be sparse but the values themselves are often larger.
3. Since the DWT W is an orthogonal matrix, this means that the wavelet transform, ϵ, of white noise, e, is itself white noise. For an explanation of this see e.g., Mardia et al. (1979). Hence, the noise is not concentrated like g but gets spread 'evenly' over all wavelet coefficients.

Using these ideas, Donoho and Johnstone (1994b) proposed that the following wavelet coefficient shrinkage technique would prove successful for estimation of $g(x)$. The idea was that large values of the empirical wavelet coefficients, d^*, were most likely cases that consisted of true signal and noise, whereas small coefficients were only due to noise. Hence, to successfully estimate d, the *thresholding* idea forms an estimate, $\hat{d}$, by removing coefficients in d^* that are smaller than some threshold and, essentially, keeps those that are larger. Donoho and Johnstone (1994b) define the hard and soft thresholding functions by

$$\hat{d} = \eta_H(d^*, \lambda) = d^* \mathbb{I}\{|d^*| > \lambda\}, \tag{3.3}$$

$$\hat{d} = \eta_S(d^*, \lambda) = \text{sgn}(d^*)(|d^*| - \lambda)\mathbb{I}\{|d^*| > \lambda\}, \tag{3.4}$$

where $\mathbb{I}$ is the indicator function, d^* is the empirical coefficient to be thresholded, and λ is the *threshold*. There are many other possibilities, for example, the *firm* shrinkage of Gao and Bruce (1997) and the Bayesian methods described in Section 3.10. A sketch of the hard and soft thresholding functions is shown in Figure 3.1. The thresholding concept was an important idea of its time introduced and applied in several fields such as statistics, approximation theory, and signal processing, see Vidakovic (1999b, p. 168).

How do we judge whether we have been successful in estimating g? Our judgement is quantified by a choice of error measure. That is, we shall define a quantity that measures the error between our estimate $\hat{g}(x)$ and the truth $g(x)$ and then attempt to choose $\hat{g}$ to try to minimize that error. The most commonly used error is the l_2 or integrated squared error (ISE) is given by

$$\hat{M} = n^{-1} \sum_{i=1}^{n} \{\hat{g}(x_i) - g(x_i)\}^2. \tag{3.5}$$

This error depends, of course, on the estimate $\hat{g}$ which depends on the particular error sequence $\{e_i\}$. We are interested in what happens with the error 'on the average' and so we define the mean ISE (MISE), or risk, by $M = \mathbb{E}(\hat{M})$. It is important to realize that M is not just a number and may depend on the estimator, the true function, the number of observations, and the properties of the noise sequence $\{e_i\}$. In wavelet shrinkage it is also especially important to remember that, since the error, M, depends on the estimator it depends not only on any 'smoothing parameters' chosen, but also on the underlying wavelet family selected to perform the smoothing (of which there are many). This important fact is sometimes overlooked by some authors.

3.3 The Oracle

Studying the risk of an estimator based on an orthogonal transform is made easier by Parseval's relation, which here says that the risk in the function domain is identical to that in the wavelet domain. Mathematically this can be expressed as

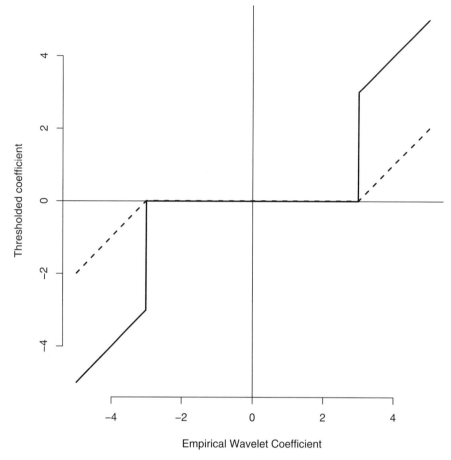

Fig. 3.1. Thresholding functions with threshold $= 3$. *Solid line*: hard thresholding, η_H; *dotted line*: soft thresholding, η_S. Produced by `f.smo1()`.

$$\hat{M} = \sum_{j,k} \left\{ \hat{d}_{j,k} - d_{j,k} \right\}^2, \tag{3.6}$$

and hence for the risk M itself. Relation (3.6) also means that we can study the risk on a coefficient-by-coefficient basis and 'get it right' for each coefficient. So for the next set of expressions we can, without loss of generality, drop the j, k subscripts.

With hard thresholding we 'keep or kill' each noisy coefficient, d^*, depending on whether it is larger than some threshold λ. The risk contribution from one coefficient is given by

$$M(\hat{d}, d) = \mathbb{E}\left\{(\hat{d} - d)^2\right\} \tag{3.7}$$

$$= \begin{cases} \mathbb{E}\left\{(d^* - d)^2\right\} \text{ if } |d^*| > \lambda \\ \mathbb{E}\left\{d^2\right\} \text{ if } |d^*| < \lambda \end{cases} \tag{3.8}$$

$$= \begin{cases} \mathbb{E}(\epsilon^2) = \sigma^2 \text{ if } |d^*| > \lambda \\ d^2 \text{ if } |d^*| < \lambda. \end{cases} \tag{3.9}$$

It is apparent from this last equation that if $d \gg \sigma$, then one would wish the first option to have held true (i.e. $|d^*| > \lambda$), which could have been 'achieved' by setting λ to be small. On the other hand, if $d \ll \sigma$, then the second option is preferable, and it would have been better to set λ to be large. Already, it can be seen that it would be preferable for the threshold to increase with increasing noise level σ. In a practical situation, which option should be picked?

To address this issue, Donoho and Johnstone (1994b) proposed the introduction of an oracle. In wavelet shrinkage the oracle is a notional device that tells you which coefficients you should select (the oracle idea itself is, as oracles tend to be, more general and its omniscience would be able to tell you, for example, where to place the knots in spline smoothing or the breaks in a piecewise constant fit). If you obey the oracle, then it always ensures that you choose the smallest of d^2 and σ^2 for each coefficient. Hence, Donoho and Johnstone (1994b) show that the *ideal* risk using the oracle is

$$M_{\text{ideal}} = \sum_{j,k} \min(|d_{j,k}|^2, \sigma^2). \tag{3.10}$$

In practice, we do not have an oracle, and the ideal risk cannot be attained in general. However, Donoho and Johnstone (1994b) established the following remarkable results:

1. If one performs wavelet shrinkage via soft thresholding with a threshold of $\sigma\sqrt{2\log n}$, then the risk of this procedure, $M_{\text{universal}}$, comes to within a log factor of the ideal risk. More precisely:

$$M_{\text{universal}} \leq (2\log n + 1)(\sigma^2 + M_{\text{ideal}}). \tag{3.11}$$

2. In terms of *ideal* risk, several established non-wavelet procedures (such as piecewise polynomial fits, variable-knot splines) are not more dramatically powerful than wavelet shrinkage (with the threshold sequence defined above). This statement means that wavelet shrinkage is never that much worse (in terms of a $\log n$ penalty) than these other methods. Of course, one needs to be careful. If the truth is exactly a piecewise polynomial then making use of piecewise polynomial fits with known break point locations will almost certainly perform better than wavelets. On the other hand, the reverse is true: a function composed of a particular kind of wavelet is best estimated using those wavelets. In practice, though, we generally do not always know the underlying function.

Donoho and Johnstone (1994b) point out that for the non-wavelet methods there is no equivalent result to (3.11). That is, there is no known theoretical result that says that for a particular practical procedure its risk is close to its ideal risk. Donoho and Johnstone (1994b) call the threshold sequence $\sqrt{2 \log n}$ the *universal* threshold, and we will say more about it in Section 3.5. They also define a set of *RiskShrink* thresholds that result in better risk performance and show that the risk obeys a similar inequality to (3.11) but with a smaller factor than $2 \log n + 1$. Bruce and Gao (1996) derive formulae for the exact bias, variance, and L_2 risk of these estimates in finite sample situations.

3.4 Test Functions

Whenever a methodology is developed, it is often the case that new testing methods are also developed in parallel (e.g. the Marron and Wand (1992) densities for testing density estimators). This is the case with wavelet shrinkage, and many test functions have been mooted. Donoho and Johnstone (1994b) introduced four such functions called Bumps, Blocks, HeaviSine, and Doppler, and Nason and Silverman (1994) introduced a particular piecewise polynomial. These functions were designed to exhibit a range of phenomena often seen in real-life data sets but often not handled well by classical nonparametric regression techniques. The Donoho and Johnstone (1994b) functions can be obtained using the `DJ.EX` function within `WaveThresh` and the piecewise polynomial using the `example.1()` function. The Donoho and Johnstone (1994b) functions are depicted in Figure 3.2 and the piecewise polynomial in Figure 3.3.

3.5 Universal Thresholding

As mentioned in Section 3.3, Donoho and Johnstone (1994b) introduced the *universal* threshold, which is given by

$$\lambda^u = \sigma \sqrt{2 \log n}. \tag{3.12}$$

In real problems the noise level σ is estimated by $\hat{\sigma}$, some estimate of the common standard deviation of the noise ϵ_i, and n is the number of observations.

As well as 'nearly' achieving the ideal risk in (3.11), the use of the λ^u threshold has another interpretation. Vidakovic (1999a) recalls the following general result by Picklands (1967).

Theorem 3.1. *Let $X_1, X_2, \ldots, X_n, \ldots$ be a stationary Gaussian process such that $\mathbb{E}X_i = 0$, $\mathbb{E}X_i^2 = 1$, and $\mathbb{E}X_i X_{i+k} = \gamma(k)$. Let $X_{(n)} = \max_{i=1}^{n}\{X_i\}$. If $\lim_{k \to \infty} \gamma(k) = 0$, then $X_{(n)}/\sqrt{2 \log n} \to 1$, almost surely, when $n \to \infty$.*

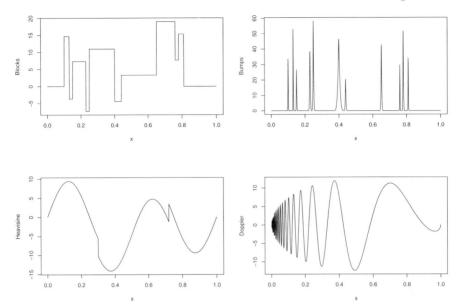

Fig. 3.2. *Top*: Blocks (*left*), Bumps (*right*); *Bottom*: HeaviSine (*left*) and Doppler (*right*) test functions introduced by Donoho and Johnstone (1994b) and drawn by the `WaveThresh` function `DJ.EX`. Produced by `f.smo2()`.

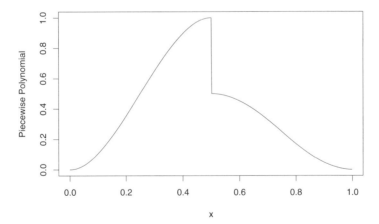

Fig. 3.3. Piecewise polynomial test function introduced by Nason and Silverman (1994) and drawn by `WaveThresh` function `example.1`. Produced by `f.smo3()`.

So, the largest of n Gaussian random variables (not necessarily independent) is roughly of the size of $\sqrt{2 \log n}$. Hence, if the wavelet coefficients were simply coefficients of Gaussian noise (of variance 1), then choosing the threshold to be $\sqrt{2 \log n}$ would eliminate the noise with high probability. For noise of different variances, the noise scale factor of σ is incorporated into λ^u. Vidakovic (1999a) also notes that if $X_1, X_2, \ldots, X_n$ are independent $N(0, 1)$, then, for large n, it can be shown that

$$\mathbb{P}(|X_{(n)}| > \sqrt{c \log n}) \approx \frac{\sqrt{2}}{n^{c/2-1}\sqrt{c\pi \log n}}. \qquad (3.13)$$

Thus, the number 2 in $\sqrt{2 \log n}$ is carefully chosen. If c in the above expression is ≤ 0, then the right-hand side tends to zero and, in wavelet shrinkage terms, this means that the largest wavelet coefficient (based on Gaussian noise) does not exceed the threshold with a high probability.

Universal thresholding can be carried out within `WaveThresh` as follows. As a running example we shall use the Bumps function. First, we shall create a noisy version of the Bumps function by adding iid $N(0, \sigma^2)$ (pseudo-)random variables. For simulation and comparing methods we usually select σ^2 to satisfy a fixed *signal-to-noise* ratio (SNR). The SNR is merely the ratio of the sample standard deviation of the signal (although it is not random) to the standard deviation of the added noise. A low SNR means high noise variance relative to the signal size.

We first generate and plot the Bumps function using the `DJ.EX` function.

```
> v <- DJ.EX()
> x <- (1:1024)/1024 # Define X coordinates too
> plot(x, v$bumps, type="l", ylab="Bumps")
```

This plot appears in the top left of Figure 3.4. Next we need to calculate the standard deviation of the Bumps function itself so that we can subsequently calculate the correct noise variance for the noise to add. We specify a SNR here of 2.

```
> ssig <- sd(v$bumps) # Bumps sd
> SNR <- 2 # Fix our SNR

# Work out what the variance of the noise is...
> sigma <- ssig/SNR

# ... and then generate it

> e <- rnorm(1024, mean=0, sd=sigma)
```

Now we are in a position to add this noise to Bumps and plot it.

```
> y <- v$bumps + e
> plot(x, y, type="l", ylab="Noisy bumps")
```

This plot appears in the top right of Figure 3.4. Next we calculate the DWT of Bumps and of the noisy Bumps signals and plot them for comparison.

```
#
# Plot wd of bumps
#
> xlv <- seq(from=0, to=1.0, by=0.2)
> bumpswd <- wd(v$bumps)
> plot(bumpswd, main="", sub="" ,
+     xlabvals=xlv*512, xlabchars=as.character(xlv),
+     xlab="x")
#
# Plot wd of noisy bumps for comparison
#
> ywd <- wd(y)
> plot(ywd, main="", sub="",
+     xlabvals=xlv*512, xlabchars=as.character(xlv),
+     xlab="x")
```

These two plots are in the bottom left and right of Figure 3.4 respectively. The bottom right plot shows the noisy wavelet coefficients (d^* in our notation from above), and the one on the left shows the 'true' coefficients, d. Even for Bumps, which has quite sharp peaks (high frequency features), its significant wavelet coefficients (bottom left) are mostly to be found in the coarse resolution levels. From the bottom right plot of Figure 3.2, even after considerable noise is added to the Bumps function, the significant 'true' coefficients stand out above the noise.

The coefficients in Figure 3.4 are all plotted according to the same scale. Sometimes a large coefficient on one scale can make coefficients on other scales very small. In Figures 3.5 and 3.6 we have used the `scaling="by.level"` argument of `plot.wd` to make the coefficients on each scale fit the available space. The default `scaling` argument for `plot.wd` is `"global"` where all the coefficients are on the same scale as in Figure 3.4. Figure 3.6 shows the noise much more clearly now, but the original Bumps wavelet coefficients still stand out prominently.

We are now in the position of a practitioner who wishes to denoise the noisy signal `y` (but, of course, since we set this example up, we know the true function; in reality we would not). Recall that the universal threshold is set to $\hat{\sigma}\sqrt{2 \log n}$. In this case, since $n = 1024$, we have $\sqrt{2 \log n} \approx 3.72$, so it remains to estimate σ. Donoho and Johnstone (1994b) suggest estimating σ by computing the median absolute deviation (MAD) of the finest-scale wavelet coefficients. (The MAD of a sequence computes the absolute values of all the differences of the data sequence from the median and then takes the median of those. MAD is computed in R by the `mad` function, which also applies a simple correction to ensure that when applied to Gaussian data it estimates the standard deviation unbiasedly.) Donoho and Johnstone (1994b) suggest

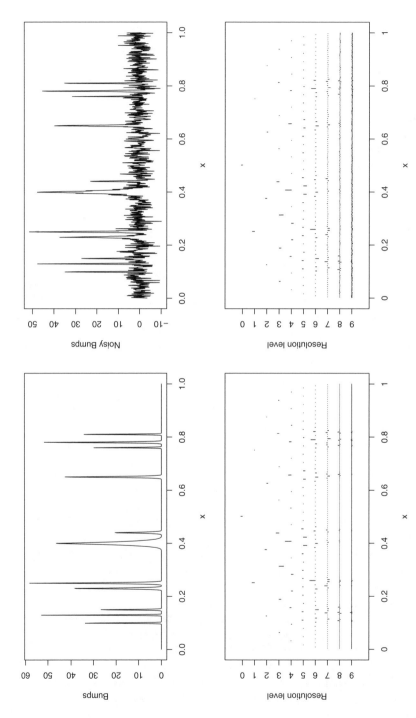

Fig. 3.4. *Top left:* Bumps function as in Figure 3.2. *Top right:* Bumps function with added iid Gaussian noise with a variance chosen to achieve a SNR of 2. *Bottom left:* DWT of Bumps. *Bottom right:* DWT of Bumps with added noise. Both wavelet transforms carried out with default WaveThresh wavelets of Daubechies' least asymmetric wavelets with ten vanishing moments. Produced by f.smo4().

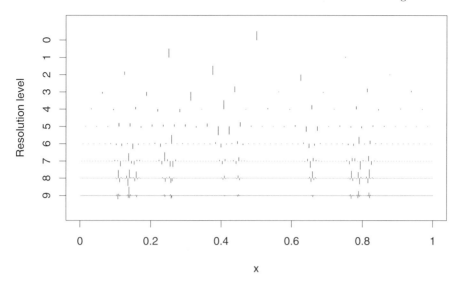

Fig. 3.5. DWT of Bumps using default `WaveThresh` wavelets of Daubechies' least-asymmetric wavelets with ten vanishing moments. Each scale level has been enlarged to fit the available width for each level. Produced by `f.smo5()`.

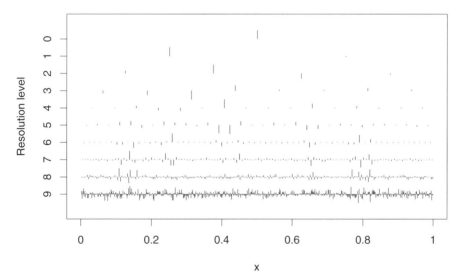

Fig. 3.6. DWT of Bumps with noise using default `WaveThresh` wavelets of Daubechies' least-asymmetric wavelets with ten vanishing moments. Each scale level has been enlarged to fit the available width for each level. Produced by `f.smo6()`.

that most true functions encountered in practice do not have much signal at
the finest-scale and hence most coefficients should just only be noise. Hence,
σ should be well estimated by just using the finest-scale coefficients. If there
were a few large signal coefficients at the finest-scale then since the MAD
estimator is robust, it should not make much difference for the estimate of σ.

In WaveThresh the threshold function performs threshold computation
and application (although there are options to (i) just compute the threshold and return it by setting return.threshold=TRUE and (ii) to apply a
previously computed threshold, by supplying a threshold as the value argument and ensuring that the policy="manual" argument is set). By default
threshold computes a threshold on, and applies it to, all coefficients in levels ranging from level three to the finest scale. In particular, it computes an
estimate of σ on all those levels. This is different to the specification given in
Donoho and Johnstone (1994b) and hence will make a difference to the performance. We show the steps necessary to compute the universal threshold using
the finest-level information only. In WaveThresh the finest-level coefficients
can be extracted by the accessD function by

```
> FineCoefs <- accessD(ywd, lev=nlevels(ywd)-1)
> sigma <- mad(FineCoefs)
> utDJ <- sigma*sqrt(2*log(1024))
```

Doing this for the ywd illustrated in Figure 3.6 gives a universal threshold of
12.44. We can threshold the wavelet coefficients at level three to the finest
scale (the default in WaveThresh, but can be changed) by

```
> ywdT <- threshold(ywd, policy="manual", value=utDJ)
```

The wd object ywdT contains the thresholded wavelet coefficients, and these
are plotted in Figure 3.7. After thresholding we are interested in knowing the
estimate expressed in the original function domain, so we take the inverse
DWT of the thresholded wavelet coefficients by

```
> ywr <- wr(ywdT)
> plot(x, ywr, type="l")
> lines(x, v$bumps, lty=2)
```

The last two plotting commands plot the denoised signal shown in Figure 3.8
as a solid line and the original Bumps signal as a dashed line. Often,
the universal threshold causes wavelet shrinkage to *oversmooth*; this is a
consequence of the *VisuShrink* 'noise-free' property. In wavelet terms this
means that too many true wavelet coefficients were deleted (or otherwise
modified), and hence too few basis functions are used to construct the
estimate. Thus, 'oversmoothing' in the wavelet case does not necessarily mean
that the estimate is 'too smooth'. If the underlying wavelet is not a smooth
function, then 'wavelet oversmoothed' estimates can look very rough.

Donoho and Johnstone (1994b) develop another set of thresholds, λ_n^*,
which satisfy a similar risk inequality to (3.11), except that $(2 \log n + 1)$ is

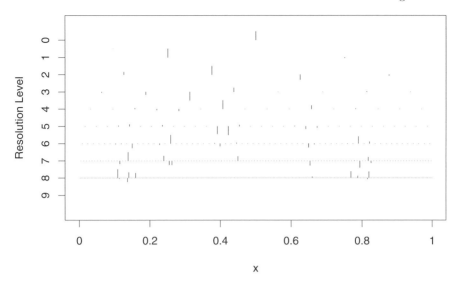

Fig. 3.7. Thresholded wavelet coefficients of noisy Bumps signal (where scaling is again performed separately for each level to show content at levels). This figure should be compared to the actual wavelet transform of the Bumps function shown in Figure 3.5. Produced by `f.smo7()`.

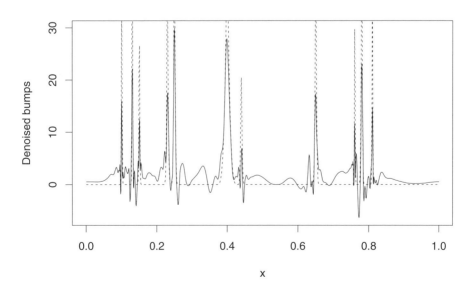

Fig. 3.8. *Solid line*: noisy Bumps signal after universal denoising. *Dashed line*: original Bumps function (which in practice we will not know). Produced by `f.smo8()`.

replaced by a smaller Λ_n^*. These thresholds form part of their *RiskShrink* procedure. For example, for the thresholding depicted in Figure 3.8 we have $n = 1024$. This means that for *VisuShrink* we have $\sqrt{2 \log n} \approx 3.72$ and $2 \log n + 1 \approx 14.86$; for *RiskShrink* we have $\lambda_n^* \approx 2.232$ and $\Lambda_n^* = 5.976$. In terms of risk, *RiskShrink* does better than *VisuShrink*.

3.6 Primary Resolution

In the previous section, the `threshold` function applied thresholding by default to resolution levels three up to the finest-scale coefficients. In practice, the success of the wavelet shrinkage for many types of threshold choice depends heavily on the choice of which resolution level to begin thresholding at. In wavelet shrinkage the coarsest level at which thresholding is applied is known as the *primary resolution*, a term coined by Hall and Patil (1995), see Section 4.7 for more details. If one examines just the scaling function part of the wavelet representation, then it has some similarities with a kernel estimator where the bandwidth h is chosen to be proportional to 2^j (see also Hall and Nason (1997)). Using just the scaling part can be interpreted as linear wavelet smoothing and is described in more detail in Section 3.11.

The success of many thresholding methods depends heavily on choosing the primary resolution correctly. However, some thresholding methods are relatively insensitive to choice of primary resolution, and these are often to be preferred, see Barber and Nason (2004) for a comparison.

3.7 SURE Thresholding

Donoho and Johnstone (1995) developed another important method for wavelet shrinkage that they called *SureShrink*. This method uses Stein's (1981) unbiased risk estimation (SURE) technique, as follows.

In the basic model given in (3.1) we assumed that the noise, e_i, was i.i.d. Gaussian with mean zero and variance of σ^2. Since the wavelet transform we consider here is orthogonal, this also means that the wavelet transform of the noise, $\epsilon_{j,k}$, is also i.i.d. Gaussian with the same mean and variance. Hence the wavelet transform of the basic model in (3.2) means that d^* is distributed as a multivariate Gaussian distribution with multivariate mean vector d and diagonal covariance matrix where all main diagonal elements are σ^2 (because of independence).

Donoho and Johnstone (1995) exploit this set-up as follows. Write the mean vector of d as $\mu = (\mu_1, \ldots, \mu_n)$ and the d^* components as independent $x_i \sim N(\mu_i, 1)$, and assume $\sigma^2 = 1$ for now. Suppose that $\hat{\mu}(\mathbf{x})$ is a particular 'nearly arbitrary, nonlinear biased' estimator for μ. Stein (1981) proposed a method for *unbiasedly* estimating the quadratic loss $||\hat{\mu} - \mu||^2$.

Stein (1981) demonstrated that if one wrote $\hat{\mu}(\mathbf{x}) = \mathbf{x} + \mathbf{g}(\mathbf{x})$, where $\mathbf{g} : \mathbb{R}^n \to \mathbb{R}^n$ was weakly differentiable, then

$$\mathbb{E}_\mu ||\hat{\mu}(\mathbf{x}) - \mu||^2 = n + \mathbb{E}_\mu \left\{ ||\mathbf{g}(\mathbf{x})||^2 + 2\nabla \cdot \mathbf{g}(\mathbf{x}) \right\}, \tag{3.14}$$

where

$$\nabla \cdot \mathbf{g} = \sum_i \frac{\partial}{\partial x_i} g_i.$$

The insight of Donoho and Johnstone (1995) was to apply Stein's result using the soft-threshold estimator given by

$$\hat{\mu}_i^{(\lambda)} = \eta_S(x_i, \lambda), \tag{3.15}$$

given in (3.4). From Figure 3.1 one can see that the partial derivative of the soft threshold with respect to x_i is given by $\mathbb{I}(|x_i| > \lambda)$ and

$$||\mathbf{g}(\mathbf{x})||^2 = \sum_{i=1}^n \hat{\mu}_i^{(\lambda)2}(\mathbf{x}) \tag{3.16}$$

$$= \sum_{i=1}^n (|x_i| - \lambda)^2 \mathbb{I}(|x_i| > \lambda), \tag{3.17}$$

as the square of the sgn function is always 1. It can be shown that the quantity

$$\mathrm{SURE}(\lambda; \mathbf{x}) = n - 2 \cdot \#\{i : |x_i| \leq \lambda\} + \sum_{i=1}^d (|x_i| \wedge \lambda)^2 \tag{3.18}$$

is therefore an unbiased estimate of the risk (in that the expectation of $\mathrm{SURE}(t; \mathbf{x})$ is equal to the expected loss). The optimal SURE threshold is the one that minimizes (3.18).

In Section 3.5 we learnt that the universal threshold is often too high for good denoising, so the minimizing value of SURE is likely to be found on the interval $[0, \sqrt{2\log n}]$. Donoho and Johnstone (1995) demonstrate that the optimal SURE threshold can be found in $\mathcal{O}(n \log n)$ computational operations, which means that the whole denoising procedure can be performed in the same order of operations.

Donoho and Johnstone (1995) note that the SURE principle does not work well in situations where the true signal coefficients are highly sparse and hence they propose a hybrid scheme called *SureShrink*, which sometimes uses the universal threshold and sometimes uses the SURE threshold. This thresholding scheme is then performed again only on certain levels above a given primary resolution. Under these conditions, *SureShrink* possesses excellent theoretical properties.

3.8 Cross-validation

Cross-validation is a well-established method for choosing 'smoothing parameters' in a wide range of statistical procedures, see, for example, Stone (1974). The usual procedure forms an estimate of the unknown function, with smoothing parameter λ based on all data except for a single observation, i, say. Then the estimator is used to predict the value of the function at i, compare it with the 'left-out' point, and then compute the error of the prediction. Then the procedure is repeated for all $i = 1, \ldots, n$ and an 'error' is obtained for the estimator using smoothing parameter λ, and this quantity is minimized over λ.

The fast wavelet transform methods in Chapter 2 require input data vectors that are of length $n = 2^J$. This fact causes a problem for the basic cross-validation algorithm as dropping a data point means that the length of the input data point is $2^J - 1$, which is no longer a power of two.

Nason (1996) made the simple suggestion of dropping not one point, but half the points of a data set to perform cross-validation. Dropping $n/2 = 2^{J-1}$ results in a data set whose length is still a power of two. The aim of the two-fold cross-validation algorithm in Nason (1996) was to find an estimate that minimizes the MISE, at least approximately. Given data from the model in Equation (3.1) where $n = 2^J$, we first remove all the odd-indexed y_i from the data set. This leaves 2^{J-1} evenly index y_i, which we can re-index from $j = 1, \ldots, 2^{J-1}$. A wavelet shrinkage estimate (with some choice of wavelet, primary resolution, hard or soft thresholding), $\hat{g}_\lambda^E$, using threshold λ, is constructed from the re-indexed y_j. This estimate is then interpolated onto the odd data positions simply by averaging adjacent even values of the estimate. In other words

$$\bar{g}_{\lambda,j}^E = (\hat{g}_{\lambda,j+1}^E + \hat{g}_{\lambda,j}^E)/2, \quad j = 1, \ldots, n/2, \tag{3.19}$$

setting $\hat{g}_{\lambda,n/2+1}^E = \hat{g}_{t,1}^E$ if g is assumed to be periodic (if it is not, then other actions can be taken). Then analogous odd-based quantities $\hat{g}^O$ and $\bar{g}^O$ are computed and the following estimate of the MISE can be computed by

$$\hat{M}(\lambda) = \sum_{j=1}^{n/2} \left\{ \left(\bar{g}_{\lambda,j}^E - y_{2j+1} \right)^2 + \left(\bar{g}_{\lambda,j}^O - y_{2j} \right)^2 \right\}. \tag{3.20}$$

The estimate $\hat{M}$ can be computed in $\mathcal{O}(n)$ time because it is based on performing two DWTs on data sets of length $n/2$. The quantity $\hat{M}(\lambda)$ is an interesting one to study theoretically: its first derivative is continuous and linearly increasing on the intervals defined by increasing $|d_{j,k}^*|$ and has a similar profile to the SURE quantity given in Section 3.7 (and could be optimized in a similar way). However, the implementation in WaveThresh uses a simple golden section search as described in Press et al. (1992).

Continuing the example from Section 3.5, we can perform cross-validated thresholding using the `policy="cv"` option of `threshold` in `WaveThresh` as follows:

```
> ywdcvT <- threshold(ywd, policy="cv", dev=madmad)
```

and then reconstruct an estimate by applying the inverse DWT

```
> ywrcv <- wr(ywdcvT)
```

and obtain the plot of the estimate given in Figure 3.9 by

```
> plot(x, ywrcv, type="l", xlab="x", ylab="Cross-val.Estimate")
> lines(x, v$bumps, lty=2)
```

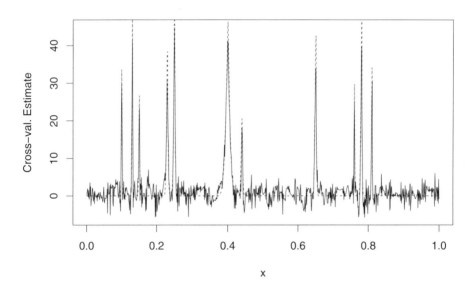

Fig. 3.9. *Solid line*: noisy bumps signal after cross-validated denoising. *Dashed line*: original Bumps function (which in practice is not known). Produced by `f.smo9()`.

The estimate in Figure 3.9 should be compared to the estimate obtained by universal thresholding in Figure 3.8. The noise-free character of the estimate in Figure 3.8 is plain to see, although the universal estimate appears to be a bit 'oversmoothed'. On the other hand, the cross-validated estimate appears to be a bit 'undersmoothed' and too much noise has been retained.

The basic cross-validation algorithm can be extended in several directions. Multivariate two-fold cross-validation was proposed by Nason (1996). Level-dependent cross-validation was proposed by Wang (1996) and generalized cross-validation by Jansen et al. (1997) and Weyrich and Warhola (1998). Nason (2002) describes an omnibus cross-validation method that chooses the 'best' threshold, primary resolution, and a good wavelet to use in the analysis.

3.9 False Discovery Rate

Abramovich and Benjamini (1996) introduced the elegant 'false discovery rate' (FDR) technology of Benjamini and Hochberg (1995) to wavelet shrinkage, see also Abramovich et al. (2006). With FDR, the problem of deciding which noisy wavelet coefficients d^* are non-zero is formulated as a multiple hypothesis testing problem. For each wavelet coefficient $d_{j,k}$ we wish to decide whether

$$H_0 : d_{j,k} = 0 \tag{3.21}$$

versus

$$H_A : d_{j,k} \neq 0, \tag{3.22}$$

for all $j = 0, \ldots, J - 1$ and $k = 0, \ldots, 2^j - 1$. If there were only one hypothesis, then it would be straightforward to implement one of several possible hypothesis tests to make a decision. In particular, one could test with a given significance level α, discover the power of the test, and so on.

However, since there are several wavelet coefficients, the problem is a multiple testing problem. It is seldom a good idea to repeat a 'single-test' significance test multiple times. For example, if there were $n = 1024$ coefficients and if $\alpha = 0.05$, then approximately $n\alpha \approx 51$ coefficients would test as positive just by chance (even if the true signal were exactly zero and $d_{j,k} = 0$ for all j, k!). In other words, many coefficients would be 'falsely discovered' as a signal.

The basic set-up of FDR as described by Abramovich and Benjamini (1996) is as follows. We assume that R is the number of coefficients that are not set to zero by some thresholding procedure. Abramovich and Benjamini (1996) then assume that of these R, S are correctly kept (i.e. there are S of the $d_{j,k}$ that are not zero) and V are erroneously kept (i.e. there are V of the $d_{j,k}$ that are kept but should not have been, because $d_{j,k} = 0$ for these). Hence $R = V + S$. They express the error in such a procedure by $Q = V/R$, which is the proportion of wrongly kept coefficients among all those that were kept. If $R = 0$, then they set $Q = 0$ (since no coefficients are kept in this case and so, of course, none can be false). The *false discovery rate of coefficients* (FDRC) is defined to be the expectation of Q. Following Benjamini and Hochberg (1995) Abramovich and Benjamini (1996) suggest maximizing the number of included coefficients but controlling the FDRC by some level q.

For wavelet shrinkage the FDRC principle works as follows (using our model notation, where m is the number of coefficients to be thresholded):

1. "For each $d^*_{j,k}$ calculate the two-sided p-value, $p_{j,k}$, testing $H_{j,k}$: $d_{j,k} = 0$,

$$p_{j,k} = 2(1 - \Phi(|d^*_{j,k}|/\sigma)).$$

2. Order the $p_{j,k}$s according to their size, $p_{(1)} \le p_{(2)} \le \cdots \le p_{(m)}$, where each of the $p_{(i)}$s corresponds to some coefficient $d_{j,k}$.
3. Let i_0 be the largest i for which $p_{(i)} \le (i/m)q$. For this i_0 calculate $\lambda_{i_0} = \sigma\Phi^{-1}(1 - p_{i_0}/2)$.
4. Threshold all coefficients at level λ_{i_0}."

(Abramovich and Benjamini, 1996, p. 5)

Benjamini and Hochberg (1995) prove that for the Gaussian noise model we assume in Equation (3.1) that the above procedure controls the FDRC at an unknown level $(m_0/m)q \le q$, where m_0 is the number of coefficients that are exactly zero. So using the above procedure will control the FDRC at a rate conservatively less than q. In practice, the method seems to work pretty well. In particular, the FDRC method appears to be fairly robust to the choice of primary resolution in that it adapts to the sparsity of the unknown true wavelet coefficients (unlike cross-validation, SURE, and the universal threshold). However, it is still the case that both the type of wavelet and a method for computing an estimate of σ are required for FDR. Recently, new work in Abramovich et al. (2006) has shown an interesting new connection between FDR and the theory of (asymptotic) minimax estimators (in that FDR is simultaneously asymptotically minimax for a wide range of loss functions and parameter spaces) and presents useful advice on the operation of FDR in real situations.

The basic FDR algorithm can be used in `WaveThresh` by using the `policy="fdr"` argument. See the help page for the `threshold` function for further details.

3.10 Bayesian Wavelet Shrinkage

Bayesian wavelet methods have always been very popular for wavelet shrinkage. The sparsity associated with wavelet representations is a kind of prior knowledge: whatever else we know (or do not know) about our function, given the earlier discussion, we usually assume that its representation will be sparse. Hence, given a set of wavelet coefficients of a deterministic function, we will know that most of them will be exactly zero, but not which ones.

A typical Bayesian wavelet shrinkage method works as follows. First, a prior distribution is specified for the 'true' wavelet coefficients, $d_{j,k}$. This prior distribution is designed to capture the sparsity inherent in wavelet representations. Then, using Bayes' theorem, the posterior distribution of the wavelet coefficients (on $d^*_{j,k}$) is computed using some, usually assumed known, distribution of the noise wavelet coefficients, $\epsilon_{j,k}$. In principle, one can calculate a posterior distribution for the unknown function by applying the inverse DWT to the wavelet coefficients' posterior distribution. However, analytically performing such a calculation is not trivial. More likely, a statistic, such as the posterior mean or median of the wavelet coefficients, is computed and then that is inverted using the inverse DWT to achieve an estimate of the 'true' function.

3.10.1 Prior mixture of Gaussians

The 'sparsity is prior knowledge' idea has been exploited by many authors. For early examples of a fully Bayesian approach see Clyde et al. (1998) and Vidakovic (1998). However, we begin our description of the Bayesian contribution to wavelet shrinkage methods with the pragmatic Chipman et al. (1997) who propose the following 'mixture of Gaussians' prior distribution for each unknown 'true' wavelet coefficient $d_{j,k}$:

$$d_{j,k}|\gamma_{j,k} \sim \gamma_{j,k}N(0, c_j^2\tau_j^2) + (1 - \gamma_{j,k})N(0, \tau_j^2), \tag{3.23}$$

where $\gamma_{j,k}$ is a Bernoulli random variable with its prior distribution of

$$\mathbb{P}(\gamma_{j,k} = 1) = 1 - \mathbb{P}(\gamma_{j,k} = 0) = p_j, \tag{3.24}$$

where p_j, c_j, and τ_j are all hyperparameters to be chosen. Model (3.23) encapsulates sparsity in the following way. The prior parameter τ_j is typically set to be small; Chipman et al. (1997) recommend that values that are inside $(-3\tau_j, 3\tau_j)$ should effectively be thought of as zero. The hyperparameter c_j^2 is set to be much larger than one. With these settings, it can be seen that the prior belief for a wavelet coefficient is that it has the possibility to be very large (distributed according to $N(0, c_j^2, \tau_j^2)$) with probability p_j. Or, with probability $1 - p_j$ it will be small (highly unlikely to be outside the $(-3\tau_j, 3\tau_j)$ interval).

Posterior distribution. One of the elegant features of Chipman et al. (1997) is that the posterior distribution is very easy to calculate. For clarity, we drop the j, k indices as they add nothing to the current exposition. Along with the priors in (3.23) and (3.24) the likelihood of the observed wavelet coefficient is

$$d^*|d \sim N(d, \sigma^2); \tag{3.25}$$

this stems from the Gaussianity assumption in the basic model (3.1). For our inference, we are interested in the posterior distribution of d given d^* denoted $(d|d^*)$. This can be derived using Bayes' theorem as follows:

$$F(d|d^*) = F(d|d^*, \gamma = 1)\mathbb{P}(\gamma = 1|d^*) + F(d|d^*, \gamma = 0)\mathbb{P}(\gamma = 0|d^*). \tag{3.26}$$

This formula can be further dissected. First, the marginal distribution of γ given d^* is

$$\mathbb{P}(\gamma = 1|d^*) = \frac{\pi(d^*|\gamma = 1)\mathbb{P}(\gamma = 1)}{\pi(d^*|\gamma = 1)\mathbb{P}(\gamma = 1) + \pi(d^*|\gamma = 0)\mathbb{P}(\gamma = 0)}$$

$$= \frac{O}{O + 1}, \tag{3.27}$$

where

$$O = \frac{\pi(d^*|\gamma = 1)\mathbb{P}(\gamma = 1)}{\pi(d^*|\gamma = 0)\mathbb{P}(\gamma = 0)} = \frac{p\pi(d^*|\gamma = 1)}{(1-p)\pi(d^*|\gamma = 0)}, \quad (3.28)$$

and $\pi(d^*|\gamma)$ is either $N(0, c^2\tau^2)$ or $N(0, \tau^2)$ depending on whether γ is one or zero respectively. Similarly $\mathbb{P}(\gamma = 0|d^*) = 1/(O+1)$.

The other conditional distributions in (3.26) can be shown to be, for $F(d|d^*, \gamma = 1)$

$$d|d^*, \gamma = 1 \sim N\left(\frac{(c\tau)^2}{\sigma^2 + (c\tau)^2}d^*, \frac{\sigma^2(c\tau)^2}{\sigma^2 + (c\tau)^2}\right) \quad (3.29)$$

and

$$d|d^*, \gamma = 0 \sim N\left(\frac{\tau^2}{\sigma^2 + \tau^2}d^*, \frac{\sigma^2\tau^2}{\sigma^2 + \tau^2}\right). \quad (3.30)$$

These two distributions are the result of the common Bayesian situation of a Gaussian prior followed by a Gaussian update d^*, see O'Hagan and Forster (2004).

Posterior mean. Chipman et al. (1997) propose using the posterior mean of d as their 'estimate' of the 'true' wavelet coefficient. Using (3.26)–(3.30) this can be shown to be

$$\mathbb{E}(d|d^*) = s(d^*)d^*, \quad (3.31)$$

where

$$s(d^*) = \frac{(c\tau)^2}{\sigma^2 + (c\tau)^2} \cdot \frac{O}{O+1} + \frac{\tau^2}{\sigma^2 + \tau^2} \cdot \frac{1}{O+1}. \quad (3.32)$$

The posterior mean of d is merely the noisy wavelet coefficient d^* but shrunk by the quantity s which can be shown to satisfy $|s| \leq 1$. Chipman et al. (1997) note that $d^*s(d^*)$ produces curves such as the ones illustrated on the left-hand side of Figure 3.10. The amazing thing about the left-hand plot in Figure 3.10 is that the function that modifies the noisy coefficient, d^*, looks very much like the thresholding function depicted in Figure 3.1: for values of d^* smaller than some critical value the posterior mean effectively sets the 'estimate' to zero, just as the thresholding functions. However, here the value is not exactly zero but very close. The solid line in Figure 3.10 corresponds to $\tau = 0.1$ and the dotted line to $\tau = 0.01$ and the 'threshold value' for the smaller τ is smaller. The posterior variance is shown in the right-hand plot of Figure 3.10, and it shows that it is most uncertain about the value of the 'true' coefficient at around the threshold value: that is, for values of d^* near to threshold value it is difficult to distinguish whether they are signal or noise.

To make use of this method one needs to obtain likely values for the hyperparameters p, c, τ, and σ. To obtain τ and $c\tau$ Chipman et al. (1997) decide what they consider to be a 'small' and 'large' coefficient by choosing reasonable values derived from the 'size' of the wavelet, and the 'size' of the function to be denoised, and from the size of a perturbation in the unknown function deemed to be negligible. For the choice of p they compute

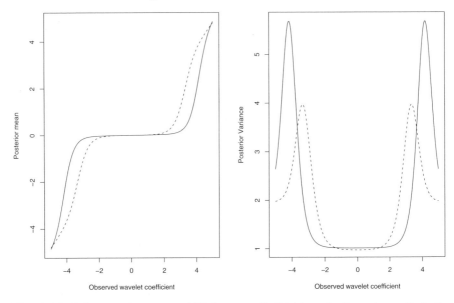

Fig. 3.10. *Left*: posterior mean, $s(d^*)d^*$ versus d^*. *Right*: posterior variance. In both plots the *solid line* corresponds to the hyperparameter choice of $p = 0.05$, $c = 500$, $\tau = 0.1$ and $\sigma = 1$. The *dotted line* corresponds to the same hyperparameters except that $\tau = 0.01$. (After Chipman et al. (1997) Figure 2). Produced by `f.smo10()`.

a proportion based on how many coefficients are larger than the universal threshold. The σ is estimated in the usual way from the data. These kinds of choices are reasonable but a little artificial. Apart from the choice τ they all relate to the noisy coefficients and are not Bayesian in the strict sense but an example of 'empirical Bayes'.

Chipman et al. (1997) were among the first to promote using the posterior variance to evaluate pointwise Bayesian posterior intervals (or 'Bayesian uncertainty bands'). We shall say more on these in Section 4.6.

3.10.2 Mixture of point mass and Gaussian

In many situations the wavelet transform of a function is truly sparse (this is the case for a piecewise smooth function with, perhaps, some jump discontinuities): that is, many coefficients are *exactly* zero and a few are non-zero. The 'mixture of Gaussians' prior in the previous section does not faithfully capture the precise sparse wavelet representation.

What is required is not a mixture of two Gaussians, but a mixture of an 'exact zero' and something else, for example, a Gaussian. Precisely this kind of mixture was proposed by Clyde et al. (1998) and Abramovich et al. (1998). Clyde et al. (1998) suggested the following prior for d:

$$d|\gamma, \sigma \sim N(0, \gamma c \sigma^2), \qquad (3.33)$$

where γ is an indicator variable that determines whether the coefficient is present (in the model) when $\gamma = 1$ or whether it is not present when $\gamma = 0$ and then the prior distribution is the degenerate distribution $N(0, 0)$.

Abramovich et al. (1998) suggest

$$d_j = \gamma_j N(0, \tau_j^2) + (1 - \gamma_j)\delta_0, \tag{3.34}$$

where δ_0 is a point mass (Dirac delta) at zero, and γ_j is defined as in (3.24). With the same likelihood model, (3.25), Abramovich et al. (1998) demonstrate that the posterior distribution of the wavelet coefficient d, given the noisy coefficient d^*, is given by

$$F(d|d^*) = \frac{1}{1+\omega}\Phi\left\{\frac{d - d^*v^2}{\sigma v}\right\} + \frac{\omega}{1+\omega}\mathbb{I}(d \geq 0), \tag{3.35}$$

where $v^2 = \tau^2(\sigma^2 + \tau^2)^{-1}$ and the posterior odds ratio for the component at zero is given by

$$\omega = \frac{1-p}{p}v^{-1}\exp\left(-\frac{d^{*2}v^2}{2\sigma^2}\right). \tag{3.36}$$

Again, we have dropped the j, k subscripts for clarity. Note that the posterior distribution has the same form as the prior, i.e., a point mass at zero and a normal distribution (note that (3.35) is the distribution function, not the density).

Posterior median. Abramovich et al. (1998) note that Chipman et al. (1997), Clyde et al. (1998), and Vidakovic (1998) all make use of the posterior mean to obtain their Bayesian estimate (to obtain a single estimate, not a whole distribution) and that the posterior mean is equivalent to a coefficient shrinkage. However, one can see from examining plots, such as that on the left hand side in Figure 3.10, that noisy coefficients smaller than some value are shrunk to a very small value. Abramovich et al. (1998) make the interesting observation that if one uses the posterior *median*, then it is actually a genuine thresholding rule in that there exists a threshold such that noisy coefficients smaller in absolute value than the threshold are set exactly to zero. For example, if $\omega \geq 1$, then this directly implies that $\omega(1+\omega)^{-1} \geq 0.5$, and as this is the coefficient of the $\mathbb{I}(d \geq 0)$ term of (3.35), the posterior distribution of $d|d^*$ has a jump discontinuity of ≥ 0.5 at $d = 0$. Here, solving for the median, $F(d|d^*) = 0.5$, would result in $d = 0$, i.e., the posterior median here would be zero (as the jump discontinuity is greater than 0.5 in size contained in a (vertical) interval of length one, and so it always must overlap the $F = 0.5$ position at $d = 0$).

3.10.3 Hyperparameters and Besov spaces.

In this book, we deliberately avoid precise mathematical descriptions of the types of functions that we might be interested in and have kept our descriptions informal. For example, the function might be very smooth (and have derivatives of all orders), or it might be continuous with jump discontinuities in its derivatives, or it might be piecewise continuous, or even piecewise constant. Statisticians are also always making assumptions of this kind. However, the assumptions are typically made in terms of statistical models rather than some absolute statement about the unknown function. For example, a linear regression model assumes that the data are going to be well represented by a straight line, whilst a global-bandwidth kernel regression estimate makes the implicit assumption that the underlying function is smooth.

One way that mathematics characterizes collections of functions is in terms of smoothness spaces. A simple example is the class of Hölder regular functions of order α, which is nicely described by Antoniadis and Gijbels (2002) as follows:

"For any $\alpha > 0$ we say that a function $f \in L^2(\mathbb{R})$ is α-Hölder regular at some point t_0 if and only if there exists a polynomial of degree n, $n \leq \alpha \leq n+1$, $P_n(t)$, and a function f_{loc} such that we may write

$$f(t_0 + t) = P_n(t) + f_{\mathrm{loc}}(t), \tag{3.37}$$

with
$$f_{\mathrm{loc}}(t) = \mathcal{O}(t^\alpha), \tag{3.38}$$

as $t \to 0$. Note that this property is satisfied when f is m-times differentiable in a neighbourhood of t_0, with $m \geq \alpha$."

One can see that the α parameter for such functions essentially provides us with a finer gradation of smoothness than integral derivatives. There are more general spaces that possess greater degrees of subtlety in function characterization. For wavelet theory, the Besov spaces are the key device for characterization. Abramovich et al. (1998) provide an accessible introduction to Besov spaces and point out that membership of a Besov space can be determined by examination of a function's wavelet coefficients (Appendix B.1.1 gives a brief explanation of this). Besov spaces are very general and contain many other spaces as special cases (for example, the Hölder space, Sobolev spaces, and other spaces suitable for representing spatially inhomogeneous functions). More information can be found in Vidakovic (1999a) and, comprehensively, in Meyer (1993b).

A major contribution of Abramovich et al. (1998) was the development of theory that links the hyperparameters of the prior, of the Bayesian model above, to the parameters of some Besov space of functions. This connection is useful for both understanding Besov spaces and for using prior Besov knowledge (or other notions of smoothness) to supply information for hyperparameter choice. Putting a prior distribution on wavelet coefficients induces a prior distribution on functions within a Besov space. Bayesian wavelet shrinkage is a type of Bayesian nonparametric regression procedure; more on Bayesian nonparametrics can be found in Ghosh and Ramamoorthi (2003).

3.10.4 Mixture of point mass and heavy tail

It is also of interest to consider other possible models for the prior of wavelet coefficients. Johnstone and Silverman (2004, 2005a,b) provide a strong case for using a mixture prior that contains a point mass at zero mixed with an observation from a heavy-tailed distribution. Heuristically, the idea behind this is that if a coefficient is zero, then it is zero(!), if it is not zero, then it has the possibility, with a heavy-tailed prior, to be large (and larger, with a high probability, than with a Gaussian component). This zero/large coefficient behaviour, and the act of finding which ones *are* large, has been coined as the statistical equivalent of 'finding a needle in a haystack'. Johnstone and Silverman (2005b) also refer to Wainwright et al. (2001), who propose that the marginal distribution of image wavelet coefficients that arise in the real world typically have heavier tails than the Gaussian.

The Johnstone and Silverman (JS) model is similar to the one in (3.34) except the Gaussian component is replaced with a heavy-tailed distribution, τ. We write their prior for a generic wavelet coefficient as

$$f_{\text{prior}}(d) = w\tau(d) + (1 - w)\delta_0(d). \tag{3.39}$$

Here, we have replaced the Bernoulli distributed γ which models the coefficient inclusion/exclusion by alternative 'mixing weight' $0 \leq w \leq 1$ (note that JS use γ for the heavy-tailed distribution; we have used τ to avoid confusion with the Bernoulli γ).

JS specify some conditions on the types of heavy-tailed distribution permitted in their theory. Essentially, τ must be symmetric, unimodal, have tails as heavy as, or heavier than, exponential but not heavier than the Cauchy distribution, and satisfy a regularity condition. The Gaussian distribution does not satisfy these conditions. JS give some examples including a quasi-Cauchy distribution and the Laplace distribution specified by

$$\tau_a(d) = \frac{a}{2} \exp(-a|d|), \tag{3.40}$$

for $d \in \mathbb{R}$ and a a positive scale parameter.

For these Bayesian methods to work well, it is essential that the hyperparameters be well chosen. JS introduce a particular innovation of 'empirical Bayes' to Bayesian wavelet shrinkage for obtaining good values for the hyperparameters. 'Empirical Bayes' is not *strictly Bayes* since parameters are estimated directly from the data using a maximum likelihood technique. However, the procedure is certainly *pragmatic* Bayes in that it seems to work well and according to Johnstone and Silverman (2004, 2005b) demonstrates excellent theoretical properties.

For example, let g be the density obtained by forming the convolution of the heavy-tailed density τ with the normal density ϕ. Another way of saying this is that g is the density of the random variable which is the sum of random variables distributed as τ and ϕ. Hence given the prior in (3.39) and the conditional distribution of the observed coefficients in (3.25), the marginal density of the 'observed' wavelet coefficients d^* is given by

$$wg(d^*) + (1 - w)\phi(d^*). \tag{3.41}$$

At this point g, ϕ and the observed wavelet coefficients are known but the w is not. So JS choose to estimate w by marginal maximum likelihood (MML). That is, they maximize the log-likelihood

$$\ell(w_j) = \sum_k \log \left\{ w_j g(d^*_{j,k}) + (1 - w_j)\phi(d^*_{j,k}) \right\}, \tag{3.42}$$

where here they estimate a separate mixing weight, $\hat{w}_j$ for each scale level (which is necessary when the noise is correlated, see Section 4.2). Then the estimated mixing weights are substituted back into the prior model and then a Bayes procedure obtains the posterior distribution. Other parameters in the prior distribution can be estimated in a similar MML fashion. The estimate of the noise variance σ is computed in the usual way by the MAD of the finest-scale wavelet coefficients, or on each level when the noise is thought to be correlated, again see Section 4.2. Further consideration on the issues surrounding this MML approach, in general, and applied to a complex-valued wavelet shrinkage can be found in Section 3.14.

Johnstone and Silverman (2005a) have made their `EbayesThresh` package available via CRAN to permit their 'empirical Bayes thresholding' techniques to be freely used. Continuing our example with the Bumps function from Section 3.8 we can use the `EbayesThresh` library to 'threshold' our noisy Bumps signal using the `ebayesthresh.wavelet` function as follows:

```
#
# Load the EbayesThresh library
#
> library("EbayesThresh")
#
# Threshold the noisy wavelet coefficients using EbayesThresh
#
> ywdEBT <- ebayesthresh.wavelet(ywd)
#
# Do the inverse transform on the shrunk coefficients
#
> ywrEB <- wr(ywdEBT)
#
# Plot the reconstruction over the original
#
> x <- (1:1024)/1024
> plot(x, ywrEB, type="l", xlab="x",
+     ylab="EBayesThresh Estimate")
> lines(x, v$bumps, lty=2)
```

The `plot` function produces an estimate as depicted in Figure 3.11. This figure should be compared to the universal threshold estimate as depicted in Figure 3.8 and the cross-validated estimate in Figure 3.9. The EbayesThresh version seems to be better than both. It is less drastically oversmoothed than the universal thresholded version and less noisy than the cross-validated estimate. EbayesThresh also has the advantage of being insensitive to primary resolution, i.e., it tends to choose its own primary resolution.

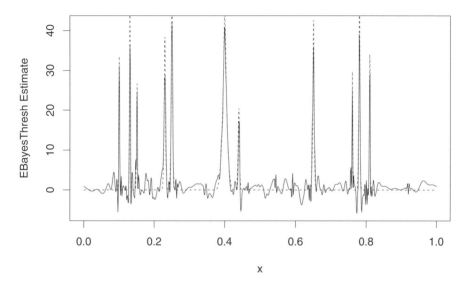

Fig. 3.11. *Solid line*: noisy Bumps signal after EbayesThresh denoising. *Dashed line*: original Bumps function. Produced by `f.smo11()`.

3.11 Linear Wavelet Smoothing

All the above wavelet shrinkage methods are *nonlinear* in the sense that the actual thresholding is not predetermined but depends on the input data. For example, in universal thresholding, we do not know what the threshold value will be until we acquire an estimate of the noise variance, σ^2 (and the number of observations). It is possible to construct a *linear* wavelet estimator by deciding *a priori* which coefficients to keep and which to kill and fixing that. Antoniadis et al. (1994) is a key reference in this area that introduces wavelet versions of some classical kernel and orthogonal series estimators.

Fixing individual coefficients in this way is not sensible: unless perhaps one knows something very specific about the signal. However, it can, and does, make sense to set to zero whole resolution levels of coefficients simultaneously. For example, if one sets to zero *all* coefficients that are finer in resolution than some scale j, then the character of the remaining function depends on the properties of the coarser wavelets up to, and including, that scale j. The fact that such an estimator is linear is easily seen as it can be constructed by merely zeroing out rows of the wavelet transform matrix itself. The 'row zeroing' and computation of the remaining wavelet coefficients can then be contained within a fixed matrix—a linear transform.

The 'finest-scale' oscillatory behaviour of the resulting estimate will be similar to the scale j wavelets. Since complete scale levels of coefficients are annihilated, the smoothing operates in a global fashion, unlike the very local behaviours of the nonlinear methods described in previous sections. WaveThresh contains the function nullevels, which is specifically designed to annihilate complete resolution levels of coefficients.

It is sometimes useful to perform linear wavelet smoothing, especially when such smoothing is part of a more complex procedure since linear smoothing is well understood theoretically. For example, Antoniadis et al. (1999) propose linear wavelet smoothing as part of a procedure to estimate a hazard rate function for survival data. Although the linear smooth is not as flexible, or potentially as effective, as a nonlinear one, it still retains the computational advantages of wavelet methods. Antoniadis et al. (1999) is further described in Section 4.8 along with an R code example of a linear wavelet smooth. A potential drawback of linear wavelet methods is that the degree of smoothing is usually not as refined as with established linear smoothers (for example, kernel regression or smoothing spline estimators). This is because established smoothers usually regulate their smoothness using a smoothing parameter on a continuum, whereas linear wavelet smoothing is regulated on a dyadic scale, see Antoniadis et al. (1994). For example, with Haar wavelets, the degree of smoothing is related to the finest-scale father wavelets, which exist on fixed scales $\dots \frac{1}{4}, \frac{1}{2}, 1, 2, 4, \dots$ etc. The comparison between smoothing on a continuum and on dyadic scales is investigated further in the wavelet context by Hall and Nason (1997).

3.12 Non-Decimated Wavelet Shrinkage

3.12.1 Translation invariant wavelet shrinkage

The non-decimated wavelet transform was introduced in Section 2.9. As mentioned there, the full non-decimated wavelet transform of a signal can be thought of as the complete set of wavelet transforms of the n cyclic shifts of the original data. Coifman and Donoho (1995) used this fact to produce a fast *translation-invariant* smoothing algorithm. Indeed, much can be achieved by averaging the results of even a few cyclic shifts. For example, one might shift the wavelet basis by 50 different amounts, perform wavelet shrinkage for each one, shift back, and then take the average of all the shift-denoise-unshift estimates. This technique is known as 'cycle spinning' and the previous example would contain 50 cycle spins. *Full* cycle spinning is where one carries out n cycle spins for a data set of length n (i.e., every shift), and it is this that forms the basis of the translation-invariant (TI) wavelet shrinkage method. The *translation invariance* means that the actual shrinkage performed does not depend on the origin of the data. In WaveThresh, the non-decimated transforms are periodic, so all shifts are cyclic.

Cycle spinning is not a wavelet shrinkage method in its own right. However, it can be used to augment, and often improve, most wavelet shrinkage techniques. Indeed, TI denoising can be thought of as a *model averaging* technique, see Clyde and George (2004) for example. That is, the wavelet shrinkage for each cycle spin is but one model that 'explains' the data, albeit at a given shift. Then, there are many models, one for each spin, and these are then averaged to get the result. In the literature to date Bayesian model averaging in wavelet shrinkage appears to have been fairly simplistic with little sophistication, such as placing priors over cycle spins. However, a nice theoretical analysis of the properties of TI denoising, including an explanation of why a higher threshold than the universal threshold $\sqrt{2 \log n}$ from Section 3.5, is required, and why the results from TI denoising are generally smoother can be found in Berkner and Wells (2002).

How would one carry out a TI denoising in WaveThresh? Let us again use our noisy Bumps example from Section 3.5. First, we need to compute the non-decimated transform of the noisy signal, y using the same (default) wavelet as previously. The following code also plots the decomposition of the noisy Bumps shown in Figure 3.12.

```
#
# Compute NDWT (packet-ordered)
#
> ywst <- wst(y)
#
# Plot it
#
> plot(ywst, main="", sub="")
```

In practice one would not have access to the true function (as that is what one is trying to estimate from the noisy data). However, since our example is a simulation, we do have access to the true Bumps function. So, let us compute the non-decimated wavelet transform of Bumps as well and plot it, as shown in Figure 3.13. Note that we have chosen to use the *packet-ordered* version of the non-decimated wavelet transform over the *time-ordered* version as the former is more suited to regression applications.

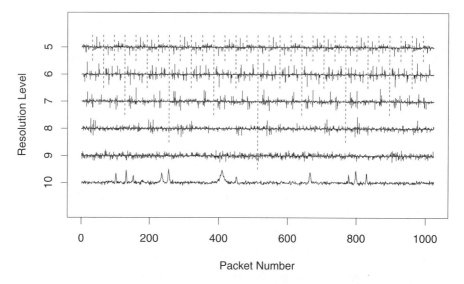

Fig. 3.12. Non-decimated wavelet transform of noisy Bumps signal. Produced by f.smo12().

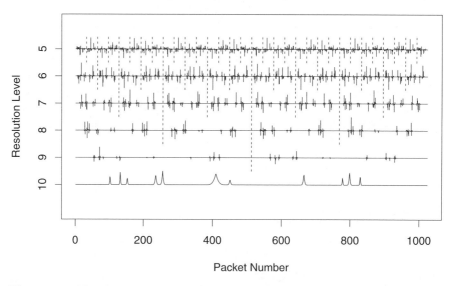

Fig. 3.13. Non-decimated wavelet transform of Bumps signal. Produced by f.smo13().

The reason one might wish to use the non-decimated transform for denoising can be seen by examining the resolution level nine coefficients in the plot of the true non-decimated coefficients in Figure 3.13. In level nine there are two packets (corresponding to a shift of zero, the left packet, and a shift of one, the right packet. Note, at this level, a shift of two would correspond again to the shift zero coefficients). In the left-hand packet, the largest coefficient corresponds to the first group of bumps in the Bumps signal. The same grouping of coefficients exists in the right-hand packet, but there is no 'largest' coefficient around this location. So the left-hand packet has possibly encoded the first set of Bumps more efficiently, and so after thresholding this large left-hand packet coefficient is likely to survive. The point is that if one used a single basis (just one of the packets), then there is a chance that the wavelets might not pick up the significant signal efficiently because of a misalignment between the wavelet basis functions and signal features. With non-decimated transforms there is 'more chance' of obtaining good alignments and hence sparse representations.

Next let us consider thresholding the noisy wavelet coefficients shown in Figure 3.12. First, let us extract the fine-scale coefficients and compute both the MAD estimator and the universal threshold (so we can compare our final TI estimator to the one in Section 3.5).

```
#
# Access the fine scale coefficients and compute
# universal threshold
#
> FineWSTCoefs <- accessD(ywst, lev=nlevels(ywd)-1)
> sigmaWST <- mad(FineWSTCoefs)
> utWSTDJ <- sigmaWST*sqrt(2*log(1024))
#
# Threshold (default number of levels) using the
# universal threshold
#
> ywstT <- threshold(ywst, policy="manual", value=utWSTDJ)

> plot(ywstT, scaling="by.level", main="", sub="")
```

Here, we compute our estimate of σ on all the fine-scale non-decimated coefficients (there are n of these, compared to when using decimated wavelets, when it would be $n/2$). Our estimate of σ in the non-decimated case is based on more wavelet coefficients, but, unlike the decimated case, the non-decimated coefficients will generally be correlated, even if the original noise is independent. We also use the same $\sqrt{2 \log n}$ universal threshold. This is because the non-decimated coefficients can be viewed as n separate bases, each one being thresholded by the usual universal threshold. Of course, this is probably not optimal since such a procedure is effectively a kind of multiple hypothesis test. A plot of the thresholded coefficients is shown in Figure 3.14 and this should be compared to the true coefficients in Figure 3.13.

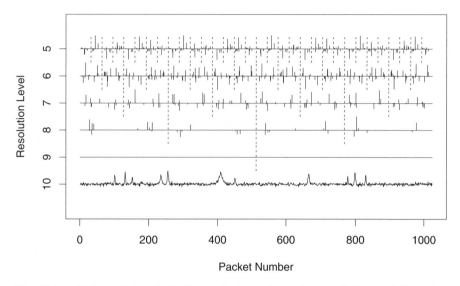

Fig. 3.14. Universal thresholded non-decimated wavelet coefficients of the noisy Bumps signal. Note that the scale 10 shows the noisy signal before thresholding. In the thresholded object (`ywstT`), only the wavelet coefficients get modified. Produced by `f.smo14()`.

The average-basis reconstruction of the thresholded coefficients can be computed using the following code:

```
> yABuv <- AvBasis(ywstT)

> yl <- range(c(yABuv, v$bumps))
> plot(x, yABuv, type="l", xlab="x",
+   ylab="TI-universal Estimate", ylim=yl)
> lines(x, v$bumps, lty=2)
```

The reconstruction, `yABuv`, is shown in Figure 3.15. Note that the TI-denoised estimate in Figure 3.15 is considerably better than that achieved with the decimated transform in Figure 3.8.

3.12.2 Basis selection

As an alternative to averaging over all shifted wavelet bases we could try to select one basis (out of all the shifted ones) that performs well. Currently, WaveThresh does not have the complete functionality to do this automatically on all signals. However, it is instructive to obtain an estimate for each shift and

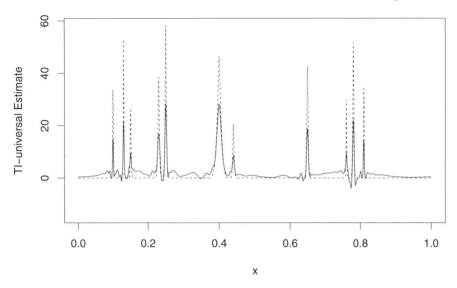

Fig. 3.15. Universal thresholded TI estimate from the noisy Bumps signal. Produced by `f.smo15()`.

then compare that estimate to the true function to get an idea of how different shifted wavelet bases can generate quite different results. In practice, though, one does not know the true function and so one has to try to select a 'best basis' by some other means (for example, some variation of the Coifman–Wickerhauser best-basis algorithm, see Coifman and Wickerhauser (1992)). Another variant would be to find not the 'best' basis but a collection of 'good' bases and then average over those.

Let us reconstruct an estimate for each shifted wavelet basis using the universally thresholded non-decimated wavelet transform computed above as ywstT.

```
#
# Create space for recording performance for
# each shifted basis. There is one shift for
# each element of y
#
> rss <- rep(0, length(y))
#
# For each shift, i, first compute the node
# vector for that shift (which defines the
# basis). Then invert ywstT using the
# packets defined by the node vector.
# Form and store the measure of performance
#
> for(i in 1:length(y))    {
```

```
+            thenv <- numtonv(i-1, nlevels(ywstT))
+            therecon <- InvBasis(ywstT, nv=thenv)
+            rss[i] <- sqrt(sum( (therecon - bumps)^2))
+            }
#
# Report performance for the standard
# wavelet basis, and the best one.
#
> cat("Standard wavelet basis RSS is ", rss[1], "\n")
> cat("Best wavelet basis RSS is ", min(rss), "\n")

#
# Plot the performances
#
> plot(1:length(y), rss, type="l",
+   xlab="Basis Number",
+   ylab="True Sum of Squares Error")
```

In the above code segment, the function InvBasis inverts a wst class non-decimated transform with respect to one of the wavelet bases specified by the nv argument. Above, the nv=thenv was supplied where the function numtonv converts a shift value into its equivalent basis description. Note also that the thresholded coefficients are only computed once and the numtonv/InvBasis combination selects each shifted basis sequentially.

Figure 3.16 shows the 'error' associated with each shifted basis (error is computed as the root sum of squares distance between the reconstruction for that basis and the 'truth'). The figure has a strong periodic appearance which occurs because the dominant Bumps features have a characteristic spacing which several subsets of wavelet bases align to in similar ways. The 'error' ranges from about 114 to 122, a difference which might be important for some applications. Although we did not mention it in the previous section, the 'error' for the basis averaging method was 111, better still. However, it is sometimes the case that a particular shifted wavelet basis can perform better than basis averaging. Note that the 'regular' wavelet basis (e.g. computed using the ordinary wd function) is the first basis in the plot of Figure 3.16. The 'regular' wavelet basis has an 'error' of 119.5, so although it was not the worst basis, it is by no means the best. All this shows how sensitive the standard wavelet method is to choice of origin and shows the need for the origin-independent methods. Finally, Figure 3.17 shows the reconstruction for the best basis (shift number 16) with an 'error' of 114 from the plot in Figure 3.16. Let us emphasize again that this basis shift was selected using complete knowledge of the true underlying function, which is not available in practice.

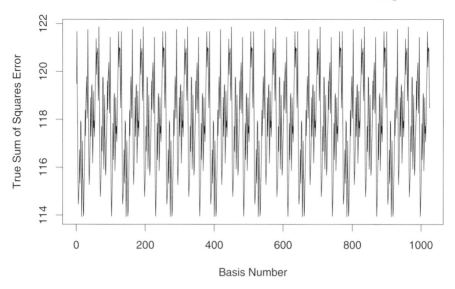

Fig. 3.16. Error of reconstruction versus basis shift. Produced by `f.smo16()`.

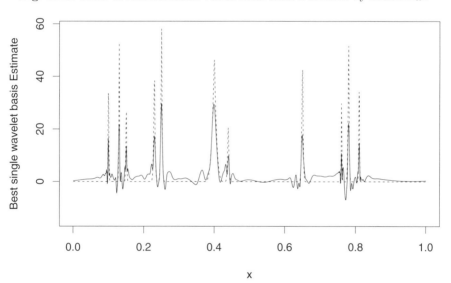

Fig. 3.17. Best reconstruction of noisy Bumps signal using basis shift 16. Produced by `f.smo17()`.

The results in Figures 3.8, 3.15, and 3.17 are all based on universal thresholding and better results might be obtained using other shrinkage methods, other wavelets, other primary resolutions, and so on. In summary:

Figure 3.8: reconstruction uses the 'regular' basis using the wd function (basis shift number 1).

Figure 3.15: reconstruction averaged over all basis shifts.

Figure 3.17: reconstruction using best-basis shift of 16.

Of all these, Figure 3.15 is best in terms of 'error' but also arguably the most visually pleasing. Figure 3.17 is next best in terms of 'error' but not as nice as Figure 3.15. Both are better than Figure 3.8.

3.13 Multiple Wavelet Shrinkage (Multiwavelets)

Multiple wavelets were described in Section 2.10, where we highlighted their potential advantages in possessing short support, orthogonality, symmetry, and vanishing moments *simultaneously*, which cannot be said for single wavelets.

Early work by Strela et al. (1999) applied the universal thresholding technology as described in Donoho and Johnstone (1994b) and Donoho (1995a). In principle, using L-dimensional multiple wavelets to denoise data is simple. One first finds a way of mapping a sequence of *univariate* noisy observations, y, to the L-dimensional input father wavelet coefficients (prefiltering), then processes these using the discrete multiwavelet transform (DMWT), performs shrinkage, then inverts the transform and maps the output vectors back to a univariate sequence (postfiltering, the inverse of prefiltering).

There are two main issues to be resolved for multiwavelet denoising discussed by Downie and Silverman (1998). The first is which prefilter to use. There are several kinds of prefilter (see Downie and Silverman (1998) for a short list and Hsung et al. (2003) for a more comprehensive discussion). However, even if the noisy sequence y is comprised of independent observations, then the multiwavelet coefficients (after prefiltering) are generally correlated. The only exception to this is the identity prefilter, where the coefficients remained uncorrelated. However, the identity prefilter is unsatisfactory for denoising for other reasons, see Downie and Silverman (1998).

The second main question is how to shrink multiwavelet coefficients? How does one threshold coefficients that are L-dimensional vectors? If the original noisy data, y, has a marginal Gaussian distribution (for example, with model (3.1)), then the multiwavelet coefficients are themselves Gaussian as they are merely a linear combination of the y. Generally, components of each vector coefficient are correlated, and different vectors of coefficients are too (although, as just mentioned, in some special cases, such as the identity prefilter, they are not). For example, Table 3.1 shows a few correlations between components within a coefficient ($\rho_{1,2}(i,0)$) and correlations between components in neighbouring coefficients (i.e. with lag, τ, equal to 1), and it can be seen that they are not negligible.

Table 3.1. Correlation between Geronimo bivariate multiwavelet coefficients using the Xia prefilter. The quantity $\rho_{a,b}(i,\tau)$ denotes the correlation between coefficient components a and b (where $a,b = 1,2$) at position i at lag τ at the finest level after taking the multiwavelet transform of 32 iid standard normal variates. Correlations are based on 5000 simulations.

i	1	2	3	4	5	6	7	8
$\rho_{1,2}(i,0)$	0.35	0.34	0.34	0.34	0.34	0.35	0.33	0.35
$\rho_{1,1}(i,1)$	−0.38	−0.40	−0.37	−0.39	−0.39	−0.38	-0.39	
$\rho_{1,2}(i,1)$	0.38	0.36	0.36	0.35	0.38	0.38	0.37	
$\rho_{2,2}(i,1)$	0.25	0.25	0.25	0.24	0.21	0.24	0.26	
$\rho_{2,1}(i,1)$	−0.26	−0.23	−0.27	−0.25	−0.26	−0.25	-0.27	

However, after taking the correlation into account we essentially still have the same set-up as in the single-wavelet case. The discrete multiwavelet transform of the signal gives us good compression and the noise is spread over the coefficients — although not in the same nice iid way as for the univariate iid case.

Downie and Silverman (1998) write the multiwavelet equivalent of univariate coefficient model (3.2) as

$$D^*_{j,k} = D_{j,k} + E_{j,k}, \tag{3.43}$$

where this time $D^*_{j,k}$, $D_{j,k}$, and $E_{j,k}$ are L-dimensional vectors, and if y is Gaussian with mean zero, then $E_{j,k}$ has the L-dimensional multivariate normal distribution $N_L(0, V_j)$. For denoising the same thresholding heuristic is applied as before. For example, for hard thresholding, the value of $D^*_{j,k}$ is assessed, for each j, k, and if it is 'large', then it is kept, and if it is 'small', it is set to zero. For multiple wavelets the question is: what does 'large' mean when $D^*_{j,k}$ is a vector?

A similar approach to that found in the single-wavelet case is taken. In the 'null' situation, where there is no signal ($D_{j,k} = 0$), the quantity $\theta_{j,k} = D^{*T}_{j,k}V_j^{-1}D^*_{j,k}$ has a χ^2_L distribution. Hence, in a practical situation the quantity $\theta_{j,k}$ can be computed and compared to an appropriate critical value. If $\theta_{j,k}$ is larger than the critical value, then $D_{j,k}$ can be estimated by $D^*_{j,k}$, otherwise the estimate will be set to zero (for hard thresholding). What is the critical value? Downie and Silverman (1998) provide similar arguments for a universal-type threshold discussed in Section 3.5. For a 'noise-free' threshold, the threshold should be conservatively set so that pure noise coefficients will be always set to zero with a high probability. So, instead of looking at the maximum of the absolute values of a sequence of Gaussian random variables, one can find a multiwavelet universal threshold by determining the maximum of a set of χ_L^2 random variables. In doing this Downie and Silverman (1998) determine their *multivariate universal threshold* by

$$\lambda_n^2 = 2\log n + (L-2)\log\log n. \tag{3.44}$$

In the single-wavelet case ($L = 1$) this formula reduces to the usual universal threshold $\lambda_n = \sqrt{2\log n}$ (which fits nicely as a χ_1^2 random variable is the square of a Gaussian). Downie and Silverman (1998) conclude that if the multivariate thresholding approach is adopted, it gives good results and certainly performs better than using a univariate threshold on each component of a multiwavelet coefficient.

Multiple wavelets are undoubtedly a good idea in that at each time-scale 'location' you appear to have more than one coefficient giving you information on the structure of the signal at that location. Although that information is sometimes somewhat dispersed due to the intercoefficient correlations. However, this author prefers complex-valued wavelet shrinkage, discussed in the next section, as a similar methodology can be developed, but a prefilter is not required. Although considerable ingenuity has gone into prefilter design, it does seem a little artificial to map univariate numbers to vectors compared to the complex-valued situation where real (univariate) numbers are *automatically* complex-valued. On the other hand, the multiwavelet system is fairly easy to extend to multiple wavelets with L greater than two, but this is not the case with complex-valued wavelets. Bui and Chen (1998) consider TI denoising with multiwavelets.

The multiple wavelet shrinkage routine in WaveThresh is threshold.mwd, which cooperates with the forward and inverse multiwavelet functions mwd and mwr as described in Section 2.10.

3.14 Complex-valued Wavelet Shrinkage

The complex-valued wavelet transform was described in Section 2.5.2. We again consider the data model (3.1) and the complex-valued wavelet transformed model is again (3.2):

$$d^* = d + \epsilon, \tag{3.45}$$

but now the wavelet coefficients d^*, d, and ϵ are all vectors of complex numbers. As with multiwavelets, the complex-valued wavelet transform takes real numbers, y, to a decomposition where there are two coefficients at each time-scale location j, k. However, whereas multiwavelets require a prefilter to map univariate numbers to a multiple wavelet input, complex-valued wavelets do not because real numbers are automatically complex (since $\mathbb{R} \subset \mathbb{C}$). In other words, the complex-valued wavelet transform directly copes with real numbers.

Several authors have considered using complex-valued wavelets for denoising. For example, Lina and MacGibbon (1997), Lina (1997), and Lina et al.

(1999) concentrate on image denoising with a Bayesian shrinkage rule. Sardy (2000) considers the estimation of complex-valued signals with universal-like thresholds (similar to the Downie–Silverman multiwavelet threshold as described in the previous section). All this work shrinks the modulus of the complex coefficients leaving the phase alone (known as *phase-preserving* shrinkage). Other work with complex wavelets in this area includes Zaroubi and Goelman (2000), which denoises complex-valued MRI scans by separate (univariate) thresholding of the real and imaginary parts, and Clonda et al. (2004) on image estimation and texture classification using a hierarchical Markov graphical model. We concentrate here on the exposition given in Barber and Nason (2004) because it concentrates on denoising a real-valued univariate function and, more to the point, the methods are implemented in WaveThresh in the subpackage cthresh.

If the input noise in model (3.1) is iid, then the individual components of the wavelet-transformed version, ϵ in (3.45), considered as *complex-valued* random variables, are uncorrelated. However, as with the multiwavelet case, the real and imaginary parts of components of vector ϵ are themselves (univariate) normal real-valued random variables, and they can be correlated. For example, Proposition 1 of Barber and Nason (2004) shows that

$$\text{cov}\left\{\Re(\epsilon), \Im(\epsilon)\right\} = -\sigma^2 \Im(WW^T)/2, \tag{3.46}$$

$$\text{cov}\left\{\Re(\epsilon), \Re(\epsilon)\right\} = \sigma^2 \left\{I_n + \Re(WW^T)/2\right\}, \tag{3.47}$$

$$\text{cov}\left\{\Im(\epsilon), \Im(\epsilon)\right\} = \sigma^2 \left\{I_n - \Re(WW^T)/2\right\}, \tag{3.48}$$

where $\Re$ and $\Im$ take the real and imaginary parts of complex numbers respectively. Figure 3.18 shows the covariance matrix $\text{cov}\left\{\Re(\epsilon), \Im(\epsilon)\right\}$ for a noise vector of $n = 128$ independent $N(0, 1)$ random variables. Covariances between a coefficient and its neighbour can be seen near the main diagonal. Also, covariances between a coefficient and its neighbour on different scales can be seen on the other diagonals. Proposition 1 from Barber and Nason (2004) shows that each complex coefficient, $d^*_{j,k}$, can be considered as a bivariate Gaussian random vector with mean zero (if $d_{j,k} = 0$) and 2×2 variance matrix of $\Sigma_{j,k}$ with entries determined by the appropriate entries from (3.46)–(3.48). As usual, the quantity σ is not known. Like the univariate case, Barber and Nason (2004) estimate σ^2 by the sum of the squared MAD of the real and imaginary parts of the finest-level coefficients.

As for thresholds and wavelet shrinkage, Barber and Nason (2004) propose using the Downie–Silverman multiwavelet threshold (CMWS) as described in Section 3.13 and, further, they prove an upper bound on the risk of hard thresholding using the multiwavelet universal threshold for the complex-valued case and draw parallels with the equivalent result for the univariate coefficients in Johnstone and Silverman (2004).

Prior specification. Barber and Nason (2004) also consider the use of Bayesian wavelet shrinkage in the complex-valued case. They abbreviate this method by CEB (complex empirical Bayes). Here, the priors on wavelet

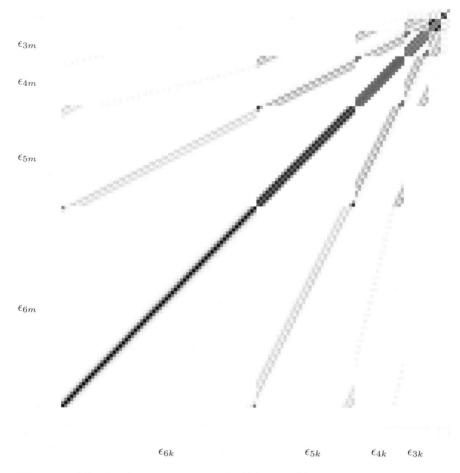

Fig. 3.18. (Absolute) covariance between $\Re(\epsilon)$ and $\Im(\epsilon)$ for a noise vector of length $n = 128$ with variance $\sigma^2 = 1$ decomposed with complex Daubechies wavelets $N = 5.5$ The *axes* correspond to wavelet coefficients d_j at levels $j = 0, \ldots, 6$. *White* corresponds to zero covariance and *black* to the maximum absolute covariance (0.29). Reproduced with permission from Barber and Nason (2004).

coefficients need to be bivariate normal so as to handle both the real and imaginary parts simultaneously. The prior they consider is of the form

$$d_{j,k} \sim w_j N_2(0, V_j) + (1 - w_j)\delta_0, \qquad (3.49)$$

where N_2 is the bivariate Gaussian distribution. Crucially, here δ_0 is the bivariate delta function at $0 + 0i$ and V_j is a 2×2 covariance matrix. Prior (3.49) is essentially the complex-valued version of the Abramovich et al. (1998) prior given in (3.34).

Posterior distribution. From model (3.45) it is clear that the likelihood for the $(j, k)th$ coefficient can be written as $d^*_{j,k} \sim N_2(d_{j,k}, \Sigma_{j,k})$. Given an observed wavelet coefficient $d^*_{j,k}$, it can be shown using standard results for the multivariate normal distribution that the posterior distribution of $d_{j,k}$ given $d^*_{j,k}$ is

$$d_{j,k}|d^*_{j,k} \sim \tilde{w}_{j,k} N_2(\mu_{j,k}, \tilde{V}_j) + (1 - \tilde{w}_{j,k})\delta_0, \qquad (3.50)$$

which is of the same form as the prior (3.49), where

$$\tilde{w}_{j,k} = \frac{w_j f(d^*_{j,k}|w_j = 1)}{w_j f(d^*_{j,k}|w_j = 1) + (1 - w_j) f(d^*_{j,k}|w_j = 0)} \qquad (3.51)$$

and

$$\tilde{V}_j = \left(V_j^{-1} + \Sigma_j^{-1}\right)^{-1} \qquad (3.52)$$

and

$$\mu_{j,k} = \tilde{V}_j \Sigma_j^{-1} d^*_{j,k}, \qquad (3.53)$$

see O'Hagan and Forster (2004) for example. The parallels to the univariate case given earlier are clear.

Choice of estimator. We could take the posterior mean as our estimator for the 'true' function (true Bayesians would probably be content to know the full posterior distribution). We do know the full posterior for the coefficients, but, due to the complexities of the wavelet transform, we do not know the complete posterior distribution of the inverse of the wavelet coefficients. The problem of obtaining posterior information on the actual function, in the original domain, given full posterior knowledge on the coefficients is addressed, for real-valued wavelets, in the section on wavelet confidence intervals in Section 4.6.

In the real-valued case the posterior *median* results in a true thresholding rule, and it has superior theoretical properties, see Johnstone and Silverman (2005b). Barber and Nason (2004) desired to use the posterior median in the complex-valued case but make the point that, in more than one dimension, there are several possibilities for the median and it is not clear which is the best or even appropriate. Further, many multivariate medians are not simple or fast to compute. So, in the event, Barber and Nason (2004) propose using three estimators: (i) the posterior mean $\tilde{w}_{j,k}\mu_{j,k}$ itself, (ii) a 'keep or kill' policy $d^*_{j,k}I(\tilde{w}_{j,k} > 1/2)$ (i.e. the observed wavelet coefficient, $d^*_{j,k}$, is kept if the posterior mixing parameter, which indicates whether the coefficient should be zero or not, and (iii) 'MeanKill', which is the same as 'keep or kill' except the posterior mean is returned instead of the observed wavelet coefficient.

Prior parameters. Barber and Nason (2004) follow Johnstone and Silverman (2005b) in using a marginal maximum likelihood method for estimating the prior covariance matrix V_j and prior mixing parameter w_j. However, the multivariate case is more tricky in that the likelihood has to be optimized over more parameters compared to the real-valued case.

Barber and Nason (2004) report an extensive simulation study and show that complex-wavelet denoising is extremely effective and outperforms anything previously encountered in this chapter, including several 'block thresholding' methods to be briefly described in Section 3.15. Barber and Nason (2004) also report that the complex-valued methods are not particularly sensitive to the choice of primary resolution, even one based on the universal multiwavelet threshold, CMWS. This suggests that it is probably not the multiwavelet transform itself, in Section 3.13, that is sensitive to primary resolution, but possibly the prefiltering action. A further interesting point is that the CMWS procedure is extremely quick to compute and apply, and it has a similar performance to the complex-valued Bayesian methods. Indeed, the CMWS was frequently faster than EbayesThresh (or complex-valued Bayesian methods); the main reason for this is that the marginal maximum likelihood optimization for the prior parameters, which is relatively time consuming.

In WaveThresh, the main function for complex-valued wavelet denoising is called cthresh. Let us continue our noisy Bumps example from Section 3.5. We will apply the cthresh function with the details=TRUE option so that we can examine several stages of the procedure. Figure 3.19 shows the complex-valued wavelet coefficients of the noisy Bumps signal, y. This figure was obtained by running the cthresh command by

```
> cmws <- cthresh(y, details=TRUE)
```

The details option ensures that many useful objects associated with the complex-valued wavelet denoising get returned. In particular, Figure 3.19 was produced by the following command:

```
> plot(cmws$data.wd)
```

Like the equivalent plots for real-valued univariate wavelets, significantly large coefficients appear at the Bumps locations. By default, cthresh uses the CMWS shrinkage. The thresholded coefficients are illustrated in Figure 3.20 and produced using the following command:

```
> plot(cmws$thr.wd)
```

After thresholding cthresh applies the inverse complex-valued wavelet transform and the resulting estimate is returned in the estimate component. Generally, the estimate component is complex-valued. Figures 3.21 and 3.22 show the real and imaginary parts of the returned estimate. These figures were produced using the following commands:

```
> yl <- range(c(Re(cmws$estimate), v$bumps))
> plot(x, Re(cmws$estimate), type="l", xlab="x",
+     ylab="Complex MW Estimate (Real)", ylim=yl)
> lines(x, v$bumps, lty=2)
```

and the same commands but replacing Re, which extracts the real part, by Im, which extracts the imaginary. The real part of the estimate in Figure 3.21

Wavelet Decomposition Coefficients

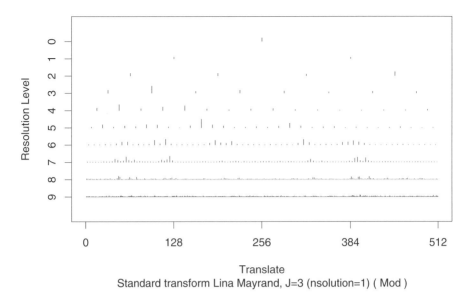

Standard transform Lina Mayrand, J=3 (nsolution=1) (Mod)

Fig. 3.19. Modulus of complex-valued wavelet coefficients of noisy Bumps signal y. Wavelet was Lina Mayrand 3.1 wavelet (also known as the Lawton wavelet). Produced by `f.smo18()`.

is pretty good. One should compare it to the estimate produced by universal thresholding in Figure 3.8, the cross-validated estimate in Figure 3.9, and the `EbayesThresh` estimate in Figure 3.11. Although the complex-valued wavelet estimate looks very good, one must be constantly careful of comparisons since the earlier estimates were based on using Daubechies' wavelets with ten vanishing moments, and the complex-valued one here relied on three. On the other hand, the extensive simulation study in Barber and Nason (2004) demonstrated that complex-valued wavelets performed well on a like-for-like basis with respect to vanishing moments.

As mentioned earlier in Section 2.10, there are often many different types of complex-valued wavelet for a given number of vanishing moments. It is possible to perform the complex-valued wavelet shrinkage for each one and average the result. Barber and Nason (2004) refer to this procedure as 'basis averaging over wavelets'. This can be achieved within `cthresh` by using an integral filter number (e.g. five would average over all the wavelets with five vanishing moments, but 5.1 would use a specific wavelet solution). Furthermore, it is also possible to perform the regular kind of basis averaging that was discussed in Section 3.12.1. This can simply be achieved using `cthresh` by setting the `TI=TRUE` option. Figures 3.23 and 3.24 show the

Wavelet Decomposition Coefficients

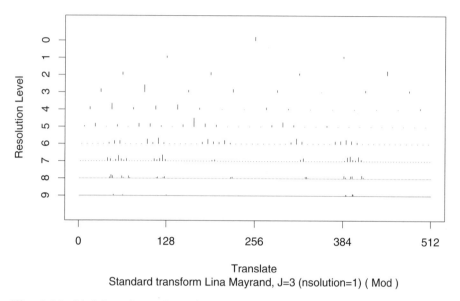

Standard transform Lina Mayrand, J=3 (nsolution=1) (Mod)

Fig. 3.20. Modulus of complex-valued wavelet coefficients from Figure 3.19 after being thresholded using CMWS thresholding. Produced by `f.smo19()`.

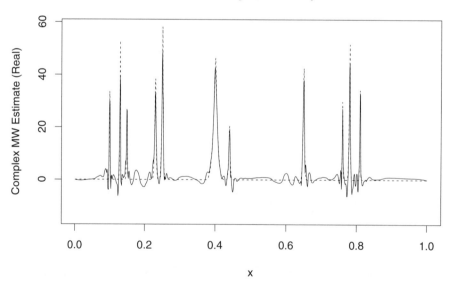

Fig. 3.21. Real part of Bumps signal estimate. Produced by `f.smo20()`.

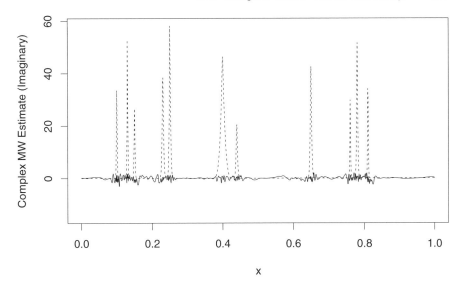

Fig. 3.22. Imaginary part of Bumps signal estimate. Produced by `f.smo21()`.

results of the translation-invariant complex-valued wavelet shrinkage using the previous `cthresh` command with the `TI` option turned on.

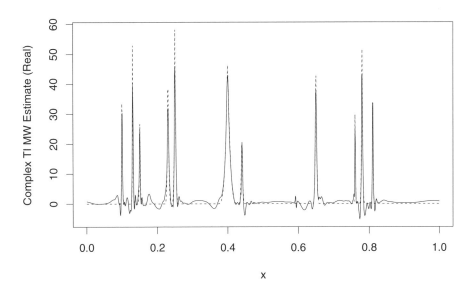

Fig. 3.23. Real part of Bumps signal TI estimate. Produced by `f.smo22()`.

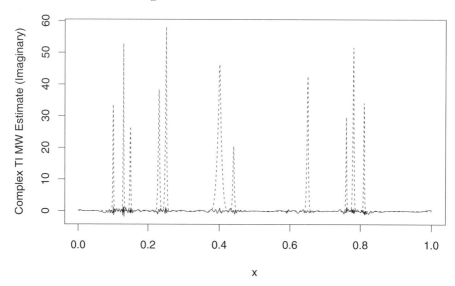

Fig. 3.24. Imaginary part of Bumps signal TI estimate. Produced by `f.smo23()`.

3.15 Block Thresholding

The general idea behind block thresholding is that one does not threshold wavelet coefficients individually, but one decides which coefficients to threshold by examining groups of coefficients together. The underlying reason for the success of block thresholding methods is that even a very 'narrow' feature in a function, such as a jump discontinuity, can result in more than one large wavelet coefficient, all located in nearby time-scale locations.

For example, construct the following simple piecewise constant function in WaveThresh by typing:

```
> x <- c(rep(0,8), rep(1,16), rep(0,8))
```

and then perform the discrete wavelet transform with $m = 4$ vanishing moments:

```
> xwd <- wd(x, filter.number=4)
```

and examine the finest-scale coefficients (rounded to three decimal places):

```
> round(accessD(xwd, level=4),3)
 [1]  0.000  0.000  0.000  0.000  0.484 -0.174  0.043
 [8]  0.000  0.000  0.000  0.000  0.000 -0.484  0.174
[15] -0.043  0.000
```

One can clearly see that the two jump discontinuities in the x function have each transformed into three non-negligible coefficients in the transform domain. So, in effect, it is the *group* of coefficients that provides evidence for some interesting behaviour in the original function.

Early work on block thresholding appeared in Hall et al. (1997) and Hall et al. (1999). Given our signal plus noise model (3.1), the (j,k)th wavelet coefficient of $g(x)$ is $d_{j,k} = \int g(x)\psi_{j,k}(x)\,dx$. Hall et al. (1997) note that $d_{j,k}$ can be estimated using the empirical quantities $\hat{d}_{j,k} = n^{-1}\sum_i y_i\psi_{j,k}(x_i)$ and that the asymptotic variance of $\hat{d}_{j,k} \approx n^{-1}\sigma^2$, even for very large (fine) values of j. To see this, note that

$$\mathrm{var}(\hat{d}_{j,k}) = n^{-2}\sum_i \sigma^2\psi_{j,k}(x_i) \approx n^{-1}\sigma^2 \int 2^j\psi(2^jx-k)^2\,dx = \sigma^2/n, \quad (3.54)$$

assuming orthonormal wavelets. Hall et al. (1997) propose to remedy the problem of 'excessive variance' of the $\hat{d}_{j,k}$ by estimating the average of $d_{j,k}$ over neighbouring k. They do this by grouping coefficients at a given scale into non-overlapping blocks of length l, with the bth block being $\mathcal{B}_b = \{k : (b-1)l + v + 1 \le k \le bl + v\}$ for $-\infty < b < \infty$ and v an integer representing an arbitrary block translation (in their numerical studies they average over all possible translations). Questions about coefficients are now transferred into questions about block quantities. So, the 'block truth', which is the average of wavelet coefficients in a block, is given by

$$B_{jb} = l^{-1}\sum_{(b)} d_{j,k}^2, \quad (3.55)$$

where $\sum_{(b)}$ means sum over $k \in \mathcal{B}_b$. The quantity B_{jb} is thought of as the approximation to $d_{j,k}^2$ for $k \in \mathcal{B}_b$. Hall et al. (1997) could estimate B_{jb} with the obvious quantity that replaces $d_{j,k}^2$ in (3.55) by $\hat{d}_{j,k}^2$, but they note that this suffers badly from bias, and so they suggest using another estimator $\hat{\gamma}_{j,k}$, which is similar to a second-order U-statistic which has good bias properties. Their overall estimator is constructed using a scaling function component (as before), but with the following block wavelet coefficient contribution:

$$\sum_{j=0}^{q}\sum_{-\infty<b<\infty}\left\{\sum_{(b)}\hat{d}_{j,k}\psi_{j,k}(x_i)\right\}\mathbb{I}(\hat{B}_{jb} > \delta^2), \quad (3.56)$$

where q is a truncation parameter that suppresses the very fine scales and δ^2 is a threshold value. The important point to note about (3.56) is that the whole set of wavelet coefficients/wavelets in block b is either totally included or totally excluded, depending wholly on whether the estimate, $\hat{B}_{jb}$, of the block truth is larger or smaller than a threshold. Hall et al. (1997) conclude that this block thresholding attains a lower bias in areas of rapidly changing signal, causes less serious 'wiggles' on the flat parts, and reduces overall MISE when compared to term-by-term thresholding. They also note that their estimator is sensitive to the truncation parameter q but appears to be relatively insensitive to the block length l.

Further investigation and development of the block thresholding approach was carried out by Cai (1999), who applied a James–Stein estimate to shrink the coefficients in each block. Cai and Silverman (2001) introduce the Neigh-Block and NeighCoeff procedures: the former estimates (shrinks) coefficients in a set of non-overlapping blocks based on information about the overall 'energy' of coefficients in a set of larger blocks, which are extensions of the smaller blocks. The larger blocks are overlapping. NeighCoeff is a special case of NeighBlock where the smaller non-overlapping blocks are individual coefficients and the larger blocks contain the coefficient and each of its immediate neighbours. Cai (2002) investigates both asymptotic and numerical aspects of block thresholding estimators and highlights the interesting fact that longer blocks are better for global adaptivity and shorter blocks are better for local adaptivity, and that the two requirements might conflict. Abramovich et al. (2002) adopt an empirical Bayes approach, as in Section 3.10.4, to block thresholding and show its superior performance to contemporary methods, and confirm the benefit of averaging over many different block origins. Chicken (2005) augments the Kovac–Silverman method for irregular data (to be described in Section 4.5) with block thresholding; more general theoretical results appear in Zhang (2005), Chicken (2007), and Chesneau (2007). Chicken and Cai (2005) deal with block thresholding for density estimation, Cai and Zhou (2008) introduce a new data-driven block thresholding technique, *SureBlock*, where block length and threshold are chosen empirically by minimization of Stein's unbiased risk estimate, see Section 3.7.

Software. At the time of writing we know of no publicly available software for R that carries out block thresholding. However, some block thresholding techniques, such as those in Cai (1999) and Cai and Silverman (2001), are available through the `GaussianWavDen` package, see Antoniadis et al. (2001)

3.16 Miscellanea and Discussion

Wavelet packet shrinkage. Various papers have addressed shrinkage and denoising using the wavelet packets described in Section 2.11. Donoho and Johnstone (1994a) provide key theoretical support to show that the 'basis-adaptive' estimator achieves, to within a log factor, the ideal loss achieved if an oracle supplied perfect information about which was the ideal basis to denoise and which coefficients were 'significant'. This significantly extends the oracle results described earlier in Section 3.3.

Discussion. Clearly, there is a huge range of wavelet shrinkage/regression techniques, and there are many that we did not cover. The obvious question to ask is what technique one should use. The answer depends on the situation at hand. For raw speed, it is difficult to beat the original *VisuShrink* and *RiskShrink* methods described in Section 3.5. The reason for the speed of these methods is, apart from calculating an estimate of σ, there is little other computation required, and the threshold can be used directly. Hence, the speed of *VisuShrink* and *RiskShrink* is governed by the speed of the forward/inverse wavelet transforms.

Antoniadis et al. (2001) conducted a large comparative simulation study of various wavelet regression methods. We refer the reader to that paper for detailed conclusions as they state that "no wavelet based denoising procedure uniformly dominates in all aspects". They also make the point that "Bayesian methods perform reasonably well at small sample sizes for relatively inhomogeneous functions, their computational cost may be a handicap, when compared with translation invariant thresholding procedures."

A recent simulation study in Barber and Nason (2004) compared complex-valued wavelet shrinkage to five other types of technique (including Bayesian block thresholding, empirical Bayesian wavelet shrinkage, multiwavelets, FDR, and cross-validation), and complex-valued methods were shown to be largely superior. Of the two complex-valued shrinkage methods, the empirical Bayesian one (CEB) gives slightly better results than the one based on the multiwavelet-type universal shrinkage (CMWS). Additionally, the CMWS method is approximately 10 times faster than `EbayesThresh` and about 80 times faster than *PostBlockMean* (block thresholding). Cai and Zhou (2008) carried out a simulation study but unfortunately only compared their *SureBlock* with the older *VisuShrink*, *SureShrink*, and *BlockJS* method of Cai (1999). A direct comparison with Barber and Nason (2004) has not been carried out, to our knowledge, but Barber (2008) reports that, by using *VisuShrink* as a benchmark, CMWS/CEB shrinkage is roughly about twice as effective in average mean-square error terms as *SureBlock* from Cai and Zhou (2008). Hence, an obvious combination would be to augment e.g., CMWS thresholding with block-thresholding technology.

Both theory and simulation studies are useful for determining which methods do well. For choice of threshold, we agree with Antoniadis et al. (2001) that empirical Bayesian methods work very well, although they can be computationally slow. The CMWS method of Barber and Nason (2004) performs almost as well as an empirical Bayesian method but is much faster.

However, a good wavelet shrinkage technique also depends on other choices, such as the identity of the wavelet underlying the wavelet transform (particularly, how many vanishing moments?). If it is thought that the underlying function is smooth (with occasional discontinuities, e.g.), then a smooth wavelet with many vanishing moments might be appropriate; for a blocky, or piecewise constant, function, then Haar wavelets might be appropriate. Of course, in many situations nothing is known about the underlying smoothness of the 'true' function, and hence our recommendation is to use wavelet methods in the first instance as they 'insure' against the presence of discontinuities and other inhomogeneities. The performance of a method can vary quite considerably depending on the underlying wavelet in the transform. Very little systematic work seems to have been performed on 'choice of wavelet' (apart from Nason (2002)), which is disappointing given its potentially dramatic effect on concrete performance.

The 'type' of the wavelet transform also is an important factor. Often translation-invariant thresholding, see Section 3.12.1, is often an effective and immediate way to improve performance. The 'cycle spinning' idea has wide application to many methods (not only wavelet shrinkage) including the regular wavelet transform, multiwavelets, block thresholding, and complex-valued shrinkage, to name but a few. We believe that both wavelet packets and lifting, described in Chapter 2, are underutilized as components of denoising algorithms. Basis averaging over different wavelets can also be a powerful tool, see Barber and Nason (2004).

4

Related Wavelet Smoothing Techniques

4.1 Introduction

The previous chapter covered a number of wavelet shrinkage techniques designed for the basic curve estimation problem described by the model given in (3.1). This chapter provides a set of complements for related problems. The first four sections explain how wavelet shrinkage might be extended to problems with correlated noise, non-Gaussian noise, multivariate data, and irregularly spaced data—all generalizing the rather restrictive modelling assumptions in (3.1). The remaining sections deal with related, but different, problems such as confidence intervals, density and survival function estimation, and inverse problems.

4.2 Correlated Data

Our first alternative to model (3.1) considers noise, e_i, $i = 1, \ldots, n$, which is not independent but correlated. Following Johnstone and Silverman (1997), suppose that the vector of noise $e = (e_1, \ldots, e_n)$ is multivariate normal with a mean vector of 0 and variance matrix of Γ. In this situation Johnstone and Silverman (1997) demonstrate that the wavelet noise vector, ϵ, has a variance matrix of $V = W\Gamma W^T$, where W is the wavelet transform matrix. Assuming stationary errors, so that the covariance values satisfy $\Gamma_{r,s} = r_{|r-s|}$, and since the filters in the wavelet transform are time-invariant (e.g., Chatfield, 2003) each level of wavelet noise coefficients $\epsilon_{j,k}$ is a stationary series as a function of k. We have neglected mentioning boundary effects here, but for more details consult Johnstone and Silverman (1997). Since $\epsilon_{j,k}$ is a stationary process, its variance can only depend on j, i.e. $\mathrm{var}(\epsilon_{j,k}) = \sigma_j^2$. See also Wang (1996). Johnstone and Silverman (1997) suggest that σ_j^2 be estimated by applying the MAD estimator, described in Section 3.5, to each level of wavelet coefficients separately. Jansen and Bultheel (1999) also describe using generalized cross-validation in a level-dependent fashion using both decimated and non-decimated wavelet transforms.

The above implies that when we take the wavelet transform of signal plus stationary correlated noise, we obtain, at each level, a sequence of wavelet coefficients with unknown mean (this is what we want to estimate) and constant variance noise. However, unlike the independent case, the wavelet coefficients of correlated data will generally be correlated. Johnstone and Silverman (1997) note the general decorrelating effect of wavelets in that correlation between wavelet coefficients at the same level tends to be less than within the original sequence and also coefficients between levels as different levels are the result of non-overlapping bandpass filters. Johnstone and Silverman (1997) illustrate both the decorrelating phenomenon and different variances at different levels with their Figures 1 and 2 using time series with a long-range correlation. We produce a similar selection of figures using an ARMA(1,1) process in Figure 4.1. Johnstone and Silverman (1997) investigate

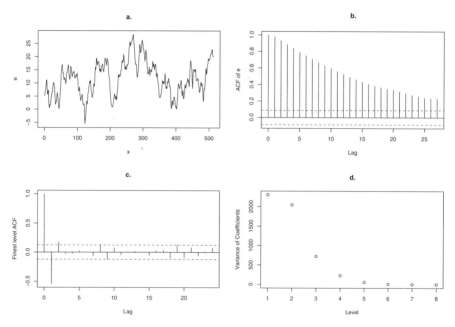

Fig. 4.1. (a) Realization of an ARIMA(1,0,1) process; (b) autocorrelation function of (a) showing significant autocorrelations even at quite high lags; (c) autocorrelations of finest-level discrete wavelet transform (using Daubechies' extremal-phase wavelet with $N = 5$ vanishing moments) demonstrating much reduced correlations; (d) variance of whole level of coefficients for each level showing different variances. Produced by f.relsmo1().

the properties of both the universal threshold (described in Section 3.5) and the SURE threshold (described in Section 3.7) for denoising in this framework.

We will use the *Bumps* example from the previous chapter, add some correlated noise, and then execute some level-dependent denoising in WaveThresh. Figure 4.2 shows the *Bumps* signal, but this time we have added highly correlated ARIMA(1,1) noise (as in Figure 4.1). The added noise has been rescaled so that the signal-to-noise ratio is two, as was the case for the independent noise in Section 3.5. Most of the large bumps are still prominent. The plot was

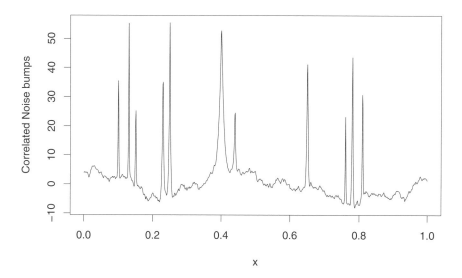

Fig. 4.2. Bumps signal with added ARIMA(1,0,1) noise so that signal-to-noise ratio is 2. Produced by f.relsmo2().

created with the following commands (assuming sigma and v are still defined as in Section 3.5):

```
#
# Generate ARMA noise
#
> eps <- arima.sim(n=1024, model=list(ar=0.99, ma=1))
#
# Create scaled noise with corrected SNR and
# then noisy signal
#
> eps <- sigma*eps/sqrt(var(eps))
> y <- v$bumps + eps
```

```
> plot(x, y, type = "l", ylab = "Correlated Noise bumps")
```

To do thresholding level by level, we just use the by.level=TRUE argument to the threshold.wd function as follows:

```
#
# Take wavelet transform
#
> ywd <- wd(y)
#
# Threshold BY LEVEL, but return the threshold
# value so we can print it out
#
> ywdT <- threshold(ywd, by.level=TRUE, policy="universal",
+    return.thresh=TRUE)
> print(ywdT)
#
# Now actually apply the threshold and invert
#
> ywr <- wr(threshold(ywd, by.level=TRUE, policy="universal"))
#
# Plot the denoised version and the original
#
> yl <- range(c(ywr, v$bumps))

> plot(x, ywr, ylim=yl, type="l")
> lines(x, v$bumps, lty=2)
```

The threshold that got printed out using the print command was

```
[1] 68.3433605 40.6977669 20.6992232 10.7933535  3.8427990
     1.3357032  0.4399269
```

The primary resolution was set to be the default, three, so the threshold, λ_j, for level j was $\lambda_3 \approx 68.3$, $\lambda_4 \approx 40.7$, and so on until $\lambda_9 \approx 0.44$. The denoised version is depicted in Figure 4.3. The reconstruction is not as good as with the universal threshold applied in Section 3.5 and depicted in Figure 3.8, but then correlated data are generally more difficult to smooth. If we repeat the above commands but using an FDR policy (i.e. replace policy="universal" with a policy="fdr"), then we get a threshold of

```
[1] 59.832912  0.000000 21.327723  9.721833  3.194402
     1.031869  0.352341
```

and the much better reconstruction as shown in Figure 4.4. As is often the case the universal thresholds were too high, resulting in a highly 'oversmoothed' estimate.

This section considered *stationary* correlated data. There is much scope for improvement in wavelet methods for this kind of data. Most of the methods

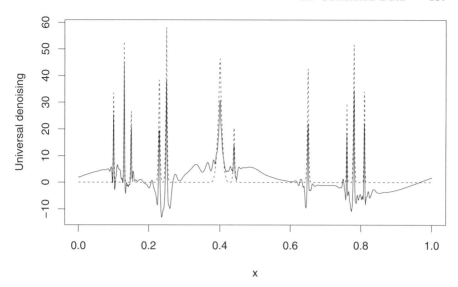

Fig. 4.3. Noisy *Bumps* denoised using universal threshold (*solid*) with true *Bumps* signal (*dotted*). Produced by `f.relsmo3()`.

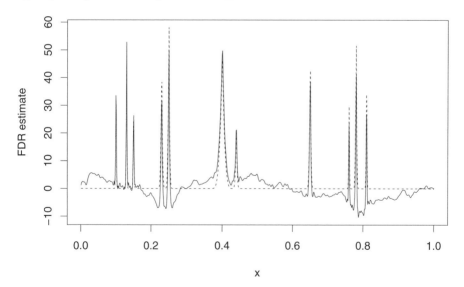

Fig. 4.4. Noisy *Bumps* denoised using levelwise FDR threshold (*solid*) with true Bumps signal (*dotted*). Produced by `f.relsmo4()`.

considered in the previous chapter were designed for iid noise and could benefit from modification and improvement for stationary correlated data. Further, and in particular for real-world data, the covariance structure might not be stationary. For example, it could be piecewise stationary or locally stationary, and different methods again would be required. The latter case is examined considered by von Sachs and MacGibbon (2000).

4.3 Non-Gaussian Noise

In many practical situations the additive noise in model (3.1) is not Gaussian but from some other distribution. This kind of problem has been considered by Neumann and von Sachs (1995) and, more recently, by Averkamp and Houdré (2003) and Houdré and Averkamp (2005).

There is also the situation of *multiplicative* non-Gaussian noise, for example, Poisson distributed noise, $X_i \sim \text{Pois}(\lambda_i)$, where the problem is to estimate the intensity sequence λ_i from the X_i. This kind of model, and other kinds of noise such as χ^2, have been considered by Donoho (1993a), Gao (1993), Kolaczyk (1997, 1999a,b), Nowak and Baraniuk (1999), Fryzlewicz and Nason (2004), and Fadili et al. (2003). More on these kinds of analysis can be found in Chapter 6.

Let us return to an example of a particular *additive* heavy-tailed noise, but restrict ourselves to analysis via Haar wavelets, making use of the ideas of Averkamp and Houdré (2003).

Suppose now that we assume that the added noise in (3.1) is such that each e_i is independently distributed as a double-exponential distribution with parameter λ (hence $\sigma^2 = \text{var}(e_i) = 2/\lambda^2$). Elementary calculations show that the characteristic function (c.f.) of the noise is

$$\chi_{e_i}(t) = \lambda^2/(\lambda^2 + t^2). \tag{4.1}$$

In the Haar wavelet transform, coarser father/mother wavelet coefficients, $c_{j-1,k}$, $d_{j-1,k}$, are obtained from finer ones by the filtering operation(s)

$$\left.\begin{array}{c} c_{j-1,k} \\ d_{j-1,k} \end{array}\right\} = (c_{j,2k} \pm c_{j,2k+1})/\sqrt{2}, \tag{4.2}$$

as described in Chapter 2. Thus, viewing the data as being at scale J, the finest-scale father and mother wavelet coefficients (scale $J-1$) have a c.f. given by

$$\chi_{c_{J-1,\cdot}}(t) = \chi_{d_{J-1,\cdot}}(t) = \left(\frac{\lambda^2}{\lambda^2 + t^2/2}\right)^2,$$

since the c.f. of the sum of two random variables is the product of their individual c.f.s (the c.f. of the difference is the same because the double exponential c.f. (4.1) is even). The Haar wavelet transform cascades the filtering operation in (4.2), and so the c.f. of any Haar father or mother wavelet coefficient at scale $j = 0, \ldots, J - 1$ is given by

$$\chi_j(t) = \left(\frac{\lambda^2}{\lambda^2 + t^2/2^{J-j}} \right)^{2^{J-j}}. \tag{4.3}$$

Further, because of the $\sqrt{2}$ in the filtering operation (4.2), the variance of each Haar father/mother wavelet coefficient at any level j remains at $\sigma^2 = 2/\lambda^2$, which can, of course, be checked by evaluating moments using (4.3). Also, the mother wavelet coefficients $d_{j,k}$ are mutually uncorrelated because of the orthonormality of the Haar discrete wavelet transform (DWT).

To simplify notation, let $m_j = 2^{J-j}$. Formula (4.3) has the same form as a Student's t-density, and, using the formula for the c.f. of this distribution from Stuart and Ord (1994, Ex. 3.13) and the duality of Fourier transforms, we can show that the Haar wavelet coefficients at level j have a density on $2m_j - 1$ degrees of freedom given by

$$f_j(x) = \frac{\lambda^*}{2^{2m_j-1}(m_j - 1)!} \exp(-\lambda^*|x|) \sum_{j=0}^{m_j-1} (2|\lambda^*|x|)^{m_j-1-j}(m_j - 1 + j)^{[2j]}/j!, \tag{4.4}$$

where $\lambda^* = \sqrt{m_j}\lambda$. It is also worth mentioning that $\lim_{j \to -\infty} \chi_j(t) = \exp(t^2/\lambda^2)$, and so the distribution of the wavelet coefficients tends to a normal $N(0, \sigma^2)$ distribution as one moves to coarser scales. Usually in wavelet shrinkage theory this fact is established by appealing to the central limit theorem as coarser coefficients are averages of the data, see e.g. Neumann and von Sachs (1995).

4.3.1 Asymptotically optimal thresholds

Theorem 2.1 of Averkamp and Houdré (2003) states that asymptotically optimal thresholds (in the sense of the previous chapter) for coefficients at level j may be obtained by finding the solutions, $\ell_{j,n}^*$, of the following equation in ℓ:

$$2(n + 1) \int_{\ell}^{\infty} (x - \ell)^2 f_j(x)\,dx = \ell^2 + \sigma^2. \tag{4.5}$$

The optimal thresholds for various values of n, for our double exponential noise, for the six finest scales of wavelet coefficients are shown in Table 4.1. The values in the table were computed by numerical integration of the integral involving $f_j(x)$ in (4.5) and then a root-finding algorithm to solve the equation. (The integration was carried out using `integrate()` in R which is based on Piessens et al. (1983); the root finder is carried out using `uniroot()`, which uses the Brent (1973) safeguarded polynomial interpolation procedure for solving a nonlinear equation.) The bottom row in Table 4.1 shows the optimal thresholds for the normal distribution (compare Figure 5

Table 4.1. Optimal thresholds, $\ell_{j,n}^*$, for Haar wavelet coefficients at resolution level j for various values of number of data points n with double exponential noise. The bottom row are the equivalent thresholds for normally distributed noise.

				n		
j	m	32	128	512	2048	65536
$J-1$	2	1.45	1.98	2.53	3.10	4.59
$J-2$	4	1.37	1.84	2.32	2.81	4.04
$J-3$	8	1.33	1.76	2.19	2.63	3.69
$J-4$	16	1.30	1.72	2.12	2.52	3.48
$J-5$	32	1.29	1.69	2.09	2.46	3.36
$J-6$	64	1.28	1.68	2.07	2.43	3.29
ϕ		*1.28*	*1.67*	*2.04*	*2.40*	*3.22*

in Averkamp and Houdré (2003)). Note how the $\ell_{j,n}^*$ converge to the optimal normal thresholds as $j \to -\infty$ and the wavelet coefficients tend to normality. Simulation studies show that the thresholds in Table 4.1 do indeed produce good results with excellent square error properties for double-exponential noise.

For further asymptotically optimal results note that the wavelet coefficients in these situations are not necessarily independent or normally distributed. However, if the data model errors, e_i, are assumed independent, and with appropriate assumptions on the function class membership of the unknown f, the usual statements about asymptotic optimality can be made (see, for example, the nice summary and results in Neumann and von Sachs (1995)).

4.4 Multidimensional Data

Wavelet shrinkage methods can be used for multidimensional problems. Most of the literature in this area is concerned with methods for image denoising: we shall briefly describe an image denoising example below. However, it is worth noting that for images (and higher-dimensional objects) that wavelets are not always the best system for representation and denoising. The main problem is that many images contain long edges which wavelets do not track sparsely. A 2D wavelet is a localized feature in all directions, and so a chain of several wavelets is usually required to accurately represent an edge. In other words, wavelet representations of images are not always sparse, see Figure 4.5 for example. Wavelets can represent some features in images sparsely, but not edges. Using the 2D wavelet transform functions described in Section 2.13 the R code to produce Figure 4.5 was

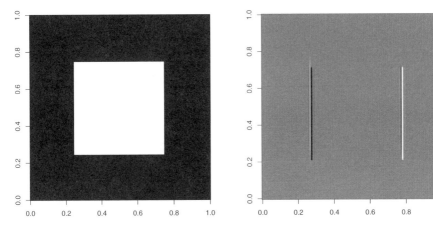

Fig. 4.5. *Left*: 256×256 image displaying a step function (*black* is zero, *white* is 1). *Right*: Finest-scale vertical wavelet coefficients of step function. Note the coefficient image is 128×128 so about 64 non-zero coefficients are required to track the single edge. This set of coefficients corresponds to the top left block from Figure 2.23. Produced by `f.relsmo5()`.

```
#
# Set-up grey scale for image colors
#
> greycol <- grey((0:255)/255)
#
# Make a box step function
#
> boxstep <- matrix(0, nrow=256, ncol=256)
> boxstep[64:(128+64-1), 64:(128+64-1)] <- 1
#
# Draw an image of the boxstep function
#
> image(boxstep, col=greycol)
#
# Perform the 2D wavelet transform on boxstep
# extract the vertical components at res
# level 7
#
> boxstepwdHORIZ7 <- matrix(imwd(boxstep)
+    [[lt.to.name(level=7, type="DC")]], nrow=128)
#
# Draw an image of the coefficients
#
> image(boxstepwdHORIZ7,  col=greycol)
```

Rather than persist with wavelets for images it is probably better to search for alternatives, and different kinds of geometric basis functions have been proposed. For example, wedgelets (Donoho, 1999; Romberg et al., 2003), beamlets (Arias–Castro et al., 2005), curvelets (Candes and Donoho, 2005a,b), and bandelets (Le Pennec and Mallat, 2005a,b). Some of these methods grow the basis functions in a given direction so that edges can be encoded extremely efficiently.

However, for completeness, if not for optimality, we demonstrate an example of image denoising. Figure 4.6 shows the original greyscale image that we use. This was obtained using the following code:

Fig. 4.6. Image of teddy bear taking his tea. Produced by `f.relsmo6()`.

```
#
# Enable access to teddy image
#
> data(teddy)
#
# Setup grey scale for image colors
#
```

```
> greycol <- grey((0:255)/255)
#
# Display image of teddy bear
#
> image(teddy, col=greycol)
```

Now add simulated Gaussian noise to this image using the following code:

```
#
# Work out 's.d.' of teddy bear
#
> bcsd <- sd(teddy)
#
# Choose a SNR of 2
#
> sdnoise <- bcsd/2
#
# Add iid noise to image
#
> noisyTeddy <- teddy + rnorm(512*512, mean=0, sd=sdnoise)
#
# Display noisy image
#
> image(noisyTeddy, col=greycol)
```

The noisy image is shown in Figure 4.7. The following code performs the 2D wavelet transform using the `imwd` function, then thresholds using the false discovery rate method but only touching the finest two levels using `threshold`, and then inverts the thresholded wavelet coefficients. The 'denoised' image appears in Figure 4.8.

```
#
# Perform WT, threshold, inverse WT & display
#
> image(imwr(threshold(imwd(noisyTeddy, filter.number=4),
+           levels=6:8, policy="fdr", dev=madmad)), col=greycol)
```

Three-dimensional wavelet shrinkage is also available within `WaveThresh` using the `wd3D` and `wr3D` forward and inverse transforms, and again shrinkage can be applied using the `threshold` function in a similar fashion to the image example given here.

4.5 Irregularly Spaced Data

One key restriction in data model (3.1) is that the data are located at equally spaced positions $x_i = i/n$. For many data sets, this regular spacing is just

Fig. 4.7. Image of teddy bear with added Gaussian noise with signal to noise ratio of 2. Produced by `f.relsmo7()`.

not realistic. Many researchers have continually sought to extend the reach of wavelet methods to data where the points x_i are distributed irregularly. The three main methods for dealing with irregularly spaced data are (i) transformations (Cai and Brown, 1998; Pensky and Vidakovic, 2001), (ii) interpolation (Hall and Turlach, 1997; Kovac and Silverman, 2000; Antoniadis and Fan, 2001; Nason, 2002), or (iii) applying statistical assumptions, such as uniform distribution, on the x_i and developing approximations (Cai and Brown, 1999; Chicken, 2003). See also Sardy et al. (1999) who develop four innovative approaches to this problem. Some recent developments involving block thresholding are Chicken (2007) and Chesneau (2007).

Below we shall briefly consider one of the above methods, mainly because it is a component of WaveThresh. However, we should mention that a new multiscale paradigm, known as *lifting*, was introduced in the mid-1990s which provides a compelling new way for handling all kinds of nonstandard data situations. In particular, lifting provides a new way for handling irregularly spaced data in a multiscale fashion. Some key publications on lifting are Sweldens (1996, 1997), and in curve estimation see Jansen et al. (2001), Delouille et al. (2004a,b). We give Nunes et al. (2006) a special mention

Fig. 4.8. Denoised image of teddy bear using wavelet shrinkage as described in the text. Produced by `f.relsmo8()`.

as not only do they introduce a new technique (adaptive lifting), but also provide a literature review on the area and refer to a complete (free) package for (adaptive) lifting in 1D called `adlift` (available from Comprehensive R Archive Network, CRAN).

As advertised, we end this section by considering the interpolation method of Kovac and Silverman (2000). Their idea was to take irregularly spaced data, i.e., where $x_i \in (0, 1)$, and interpolate the values y_i to a particular prespecified regular grid. Then the usual wavelet shrinkage is applied to the interpolated values on the regular grid with special treatment for the thresholding of the wavelet coefficients (because they are the coefficients of correlated interpolated function values, not the assumed independent function values themselves).

More precisely, Kovac and Silverman (2000) choose a new equally spaced grid $t_0, \ldots, t_{N-1}$ on $(0, 1)$, where $N = 2^J$ and $J \in \mathbb{N}$. They propose $t_k = (k + 0.5)/N$ for $k = 0, \ldots, N - 1$ and choose $N = 2^J$ such that $J = \min\{j \in \mathbb{Z} : 2^j > n\}$. Throughout they linearly interpolate the original data values y_i on x_i to new data values s_k on the new grid by

$$s_k = \begin{cases} y_1 & \text{if } t_k \leq x_1, \\ y_i + (t_k - x_i)\frac{y_{i+1}-y_i}{x_{i+1}-x_i} & \text{if } x_i \leq t_k \leq x_{i+1}, \\ y_n & \text{if } t_k > x_n. \end{cases} \qquad (4.6)$$

Kovac and Silverman (2000) note that this interpolation can be written in matrix form by

$$s = Ry, \qquad (4.7)$$

where the interpolation matrix R depends on both t (the new grid) and x, the old data locations. As mentioned above, wavelet shrinkage is applied to the new interpolated values. The first step of the shrinkage is to take the discrete wavelet transform, which we can write here in matrix form as

$$d^* = Ws, \qquad (4.8)$$

where W is the $N \times N$ orthogonal matrix associated with, say, a Daubechies orthogonal wavelet as in Section 2.1.3. If the original data observations y_i are iid with variance σ^2, it is easy to show that the variance matrix of the interpolated data, s, is given by

$$\Sigma_S = \sigma^2 R R^T. \qquad (4.9)$$

Previously mentioned wavelet shrinkage techniques (such as universal thresholding and SURE) can be modified to take account of the different variances of the wavelet coefficients. An innovation of Kovac and Silverman (2000) is a fast algorithm, based on the fast wavelet transform, to compute the variance matrix in (4.9). Indeed, from the structure of the interpolation in (4.6) it can be seen that Σ_S is a band matrix and hence further computational efficiencies can be obtained.

We now show how to use the Kovac and Silverman (2000) methodology from within WaveThresh. Figure 4.9 shows the famous motorcycle crash data taken from Silverman (1985) (this set is accessible in R after accessing the MASS library using the call library("MASS")). The 'Time' values are not regularly spaced, and so this is a suitable data set on which to exercise the irregularly spaced methods. Figure 4.9 also shows the interpolated data, s, from (4.7). The figure was produced with the following commands:

```
> library("MASS") # Where the mcycle data lives

> plot(mcycle[,1], mcycle[,2], xlab="Time (ms)",
+ ylab="Acceleration")

> Time <- mcycle[,1]
> Accel <- mcycle[,2]
#
# Rescale Time to [0,1]
#
```

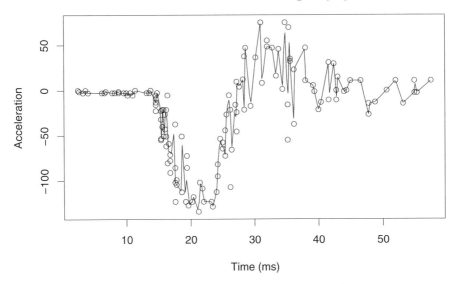

Fig. 4.9. *Points*: motorcycle crash data from Silverman (1985). *Line*: the interpolated data. Produced by `f.relsmo9()`.

```
> Time01 <- (Time - min(Time))/(max(Time) - min(Time))
#
# Interpolate data to grid
#
> McycleGrid <- makegrid(t=Time01, y=Accel)
#
# Scale new [0,1] grid back to original scale
#
> TimeGrid<-McycleGrid$gridt*(max(Time)-min(Time))+min(Time)
#
# Plot interpolated data
#
> lines(TimeGrid, McycleGrid$gridy)
```

The `makegrid` function performs the interpolation of the old data values to the new regularly spaced grid. The `makegrid` function always expects x values to lie in the interval $[0, 1]$. This explains the necessity of creating the `Time01` object above which is a linear rescaling of the `Time` values onto the interval $[0, 1]$. The return value of `makegrid` contains (among other things) two components called `gridt` and `gridy` which correspond to the new gridded values t and y as denoted above.

We can then take the DWT of the interpolated data: we could apply the regular DWT, `wd`, in `WaveThresh`. However, there is a special function called `irregwd` which performs both the DWT and the computations required to obtain the wavelet coefficient variances. The `irregwd` function can be applied as follows:

```
#
# Perform KS00 irregular wavelet transform
#
> McycleIRRWD <- irregwd(McycleGrid)
#
# Convert the irregwd object to wd for coef plotting
#
> McycleIRRWD2 <- McycleIRRWD
> class(McycleIRRWD2) <- "wd"
> plot(McycleIRRWD2)
```

and the wavelet coefficients are plotted in Figure 4.10. The other point to note is that the class of the object returned by `irregwd` is `irregwd`, and hence the `plot.irregwd` function is dispatched to plot the object. However, this function actually plots the variance factors (shown in Figure 4.11; note that since the coefficients in this figure are variances, they are all positive). To obtain the actual coefficients, the `irregwd` object has to be coerced back into a `wd` object and then plotted (as above).

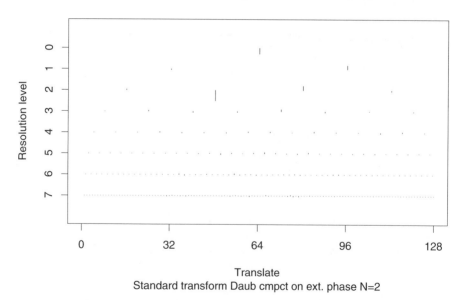

Fig. 4.10. Wavelet coefficients of interpolated data. Produced by `f.relsmo10()`.

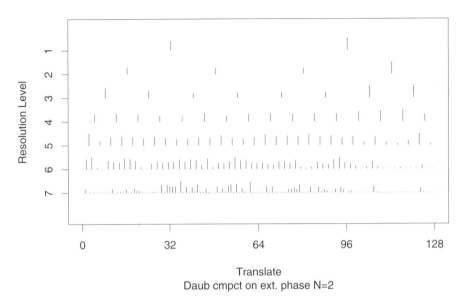

Fig. 4.11. Variance factors associated with interpolated data. Produced by
f.relsmo11().

After transforming with `irregwd` the next step in the curve estimation
procedure is to threshold the coefficients. The Kovac and Silverman (2000)
method normalizes the wavelet coefficients by their computed standard deviation and then applies standard thresholding. For example:

```
#
# Do thresholding
#
> McycleT <- threshold(McycleIRRWD, policy="universal",
+    type="soft", dev=madmad)
#
# Invert and plot, and original
#
> McycleWR <- wr(McycleT)
> plot(TimeGrid, McycleWR, type="l", xlab="Time (ms)",
+    ylab="Accel")
> points(Time, Accel)
```

After this thresholding we obtain the estimate shown in Figure 4.12. The
estimate is not extremely bad but could no doubt be improved by a judicious
choice of threshold method and parameters.

Although the Kovac and Silverman (2000) method is presented here as a
technique for analyzing irregularly spaced data, it is also a method that can

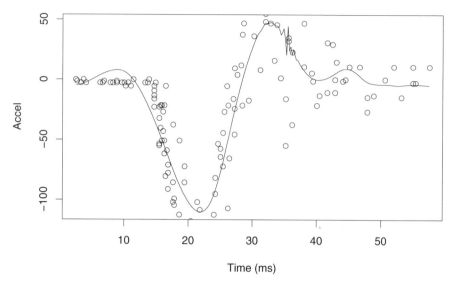

Fig. 4.12. *Line*: basic `irregwd` smooth; *points*: `mcycle` data. Produced by `f.relsmo12()`.

handle data which have a heterogeneous variance structure. For example, we could replace the identically distributed y_i with a set of data whose variance σ_i^2 varied with i. As long as a means for supplying the estimated variance to the Kovac and Silverman (2000) algorithm can be found, thresholding can be performed. See, for example, Section 7.1 of Kovac and Silverman (2000); a similar technique is used in Nunes et al. (2006).

4.6 Confidence Bands

Much less attention has been paid in the literature to the problem of confidence or credible bands for wavelet shrinkage. Can we find an interval in which we think the true function lies (usually tightened up with some probability statement about how likely the inclusion is)?

One of the earliest methods for deriving confidence intervals for wavelet shrinkage estimates was introduced by Bruce and Gao (1996) with their pointwise estimates of variance.

An early Bayesian method for computing credible intervals for wavelet shrinkage was developed by Chipman et al. (1997). As described in Section 3.10.1, they form credible intervals by using the posterior mean estimation of the function at point i, $\hat{g}_i$ (this is just the inverse wavelet transform of the posterior mean of the wavelet coefficients) and then, using the posterior variance of $[d_{j,k}|d_{j,k}^*]$, they compute the posterior variance of $\hat{g}_i$. Their 'uncertainty bands' are formed by $\hat{g}_i \pm 3\text{s.d.}\{\hat{g}_i\}$.

The intervals specified by Chipman et al. (1997) are undoubtedly simple, pragmatic, and fast to compute. However, they are often not suitable, especially with some other Bayesian models. For example, Barber et al. (2002) develop Bayesian credible intervals to work with the 'mixture of a point mass with a Gaussian' prior of Abramovich et al. (1998) as described in Section 3.10.2. The posterior distribution of the 'unknown' function can be represented as the convolution of the posterior distributions of the wavelet coefficients and the scaling coefficient:

$$[g|y] = [c_{0,0}|c_{0,0}^*]\phi(t_i) + \sum_j \sum_k [d_{j,k}|d_{j,k}^*]\psi_{j,k}(t_i), \qquad (4.10)$$

where $c_{0,0}$ is the scaling function coefficient. Barber et al. (2002) note that the convolution given in (4.10) is impractical to evaluate analytically and that direct simulation to establish the posterior of g, although possible, would be time consuming. Their approach is to estimate the first four cumulants of (4.10) and then fit a parametric distribution which matches those cumulants.

Cumulants. Suppose that the moment-generating function of a random variable X is written $M_X(t)$. The *cumulant-generating function* is $K_X(t) = \log M_X(t)$. Barber et al. (2002) write $\kappa_r(X)$ for the rth cumulant of X, which is given by the rth derivative of $K_X(t)$ evaluated at $t = 0$. It is well known that $\kappa_1(X)$ and $\kappa_2(X)$ are the mean and variance of X, $\kappa_3(X)/\kappa_2^{3/2}(X)$ is the skewness, and $\kappa_4(X)/\kappa_2^2(X) + 3$ is the kurtosis and that $\kappa_3(X)$ and $\kappa_4(X)$ are zero if X is Gaussian. For finding the cumulants of (4.10) the most important property is that if X and Y are independent, and a, b are real constants, then

$$\kappa_r(aX + b) = \begin{cases} a\kappa_1(X) + b, & r = 1, \\ a^r\kappa_r(X), & r = 2, 3, \ldots \end{cases} \qquad (4.11)$$

and

$$\kappa_r(X + Y) = \kappa_r(X) + \kappa_r(Y), \qquad (4.12)$$

for all r. More on cumulants can be found in Barndorff-Nielsen and Cox (1989) or Stuart and Ord (1994).

Applying (4.11) and (4.12) to (4.10) gives cumulants of $[g_i|y]$ as

$$\kappa_r(g_i|y) = \kappa_r(c_{0,0}|c_{0,0}^*)\phi^r(t_i) + \sum_j \sum_k \kappa_r(d_{j,k}|d_{j,k}^*)\psi_{j,k}^r(t_i). \qquad (4.13)$$

The cumulants of the wavelet and scaling function coefficients ($\kappa_r(c_{0,0}|c_{0,0}^*)$ and $\kappa_r(d_{j,k}|d_{j,k}^*)$) are easy to find from the moments of the posterior distribution given by (3.35).

The representation for the cumulants of $[g|y]$ is reminiscent of the inverse wavelet transform for $[g|y]$ itself in (4.10) *except* for one crucial difference: the scaling function, ϕ, and wavelets, $\psi_{j,k}$, have been replaced by their rth power. This is a consequence of (4.11). Generally, there is no magic formula for evaluating powers of wavelets. However, good approximations can be found, as described next.

4.6.1 Approximate powers of wavelets

For Haar wavelets defined in Section 2.2 a moment's thought reveals that the square of the Haar wavelet is just the Haar scaling function, i.e., $\psi^2(t) = \phi(t)$, and that $\psi^3(t) = \psi(t)$ and $\psi^4(t) = \phi(t)$. Hence, the dilated versions of these powers can also be easily expressed in terms of the original ψ and ϕ functions. For example,

$$\psi_{j,k}(t)^2 = \{2^{j/2}\psi(2^j t - k)\}^2 = 2^j \psi^2(2^j t - k) = 2^j \phi(2^j t - k) = 2^{j/2}\phi_{j,k}(t). \tag{4.14}$$

The key point is that if one is looking for coefficients of $\psi^2_{j,k}$, as we are in (4.13) for example, then one only need use a rescaled version of the father wavelet coefficients. These coefficients are computed automatically during the DWT (although they are not always stored. This is because the most efficient DWT operates 'in-place' with new wavelet coefficients 'overwriting' scaling function coefficients. However, the algorithm can easily be changed so as to keep the coefficients). In other words, the regular Haar DWT already and automatically contains the basic coefficients required for any wavelet power. This means that (4.13) can be computed as rapidly as the DWT itself for Haar wavelets (up to a constant factor).

The relation $\psi(t)^2 = \phi(t)$, and the others above for other powers, does not work for any other Daubechies' wavelet. However, following Herrick (2000), Barber et al. (2002) considered approximating the general wavelet $\psi^r_{j_0,0}$ for some $0 \le j_0 \le J - m$ by

$$\psi^r_{j_0,0}(t) \approx \sum_{\ell} e_{j_0+m,\ell}\phi_{j_0+m,\ell}(t), \tag{4.15}$$

where m is a positive integer. The idea is to approximate ψ^r at scale j_0 by scaling functions at the finer-scale $m_0 = j_0 + m$. The idea behind this is that we can obtain (approximate) values of the coefficients of a representation involving ψ^r from the coefficients of scaling function coefficients at finer scales. The success of these approximations is illustrated in Figure 4.13. The left-hand column of Figure 4.13 shows approximations (dotted lines) for the square of a wavelet (solid line). The top shows the approximation using scaling functions at one finer scale, the middle using scaling functions at two finer scales, and the bottom using scaling functions from three finer scales. Indeed, the bottom approximation using scaling functions from scales three levels finer than j_0 is almost indistinguishable from the true square. Similar behaviour is observed for the cube and fourth power in the middle and right-hand columns of Figure 4.13. Barber et al. (2002) find the case $m = 3$ a sufficiently accurate approximation for their purposes. The above approximation was advertised

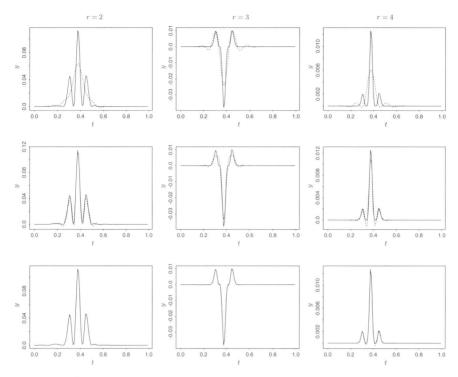

Fig. 4.13. Approximation to powers of Daubechies' least-asymmetric wavelet with eight vanishing moments; the powers are indicated at the *top* of each *column*. *Solid lines* are wavelet powers and *dotted* lines show approximations using scaling functions at level m_0. From *top* to *bottom*, graphs show approximation at levels $m_0 = 4$, $m_0 = 5$, and $m_0 = 6$; the original wavelet is at level $j_0 = 3$. Reproduced from Barber et al. (2002) with permission.

for powers of wavelets with $j_0 \leq J - m$, similar approximative arrangements can be made for $J - m < j_0 < J - 1$, the finest-scale coefficients.

Using this approximation it is possible to compute, with high accuracy, the coefficients of powers of any Daubechies' wavelet. This approximation allows us to find the first four cumulants of $[g_i|y]$ from the cumulants of the posterior distribution of the wavelet coefficients using Formula (4.13). Once the first four cumulants of the posterior distribution of the data g_i are known, the Barber et al. (2002) approximate the posterior distribution of $[g_i|y]$ by using Johnson curves, which effectively fits a parametric form to the posterior distribution, see Johnson (1949). Confidence intervals can then be obtained from the quantiles of the distributions fitted by the Johnson curves. This method is codified within WaveThresh in the wave.band function. Figure 4.14 shows an example of wave.band in action.

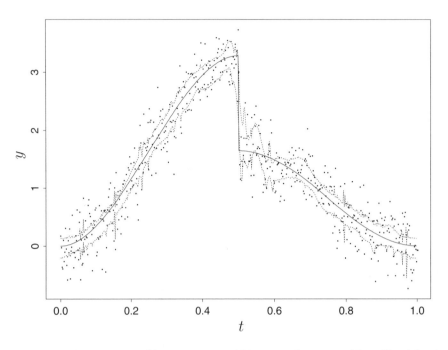

Fig. 4.14. A pointwise 99% `wave.band` credible interval estimate (*dotted line*) for the piecewise polynomial (*solid line*). *Dots*: noisy data on $n = 512$ equally spaced points with the addition of iid Gaussian noise with mean zero and SNR of 3. Reproduced with permission from Barber et al. (2002).

4.6.2 More accurate bands

Semandeni et al. (2004) describe an improved method for obtaining Bayesian credible bands again building on the Abramovich et al. (1998) work. Their method is based on using a saddlepoint approximation to the posterior distribution of the coefficients. Further, they directly compute the powers of wavelets and approximate them as above. Semandeni et al. (2004) have produced the called `SBand` package, which we have used in following example, which uses the Bumps signal as the 'truth', generates some noisy data, and then generates estimates and credible intervals.

```
> p.sig.name <- "bumps"
> p.rep <- 1
> p.n <- 1024

> p.sig.gen <- my.fct.generation(fg.sig.name = p.sig.name,
+     fg.n = p.n, fg.rep = p.rep, fg.rsnr=2)
> p.data <- p.sig.gen$data
```

```
> p.signal.y <- p.sig.gen$y


> p.example <- my.fct.SBand(my.data = p.data,
+     my.signal.y = p.signal.y)

> plot(p.example)
```

Their function `my.fct.generation` generates one of the Donoho and Johnstone test signals with Gaussian noise with a given SNR ratio (it can generate many replicates, although we only generate one here). Then, for each replicate, the function `my.fct.SBand` computes the confidence interval. In this example, we supply the true function to `my.fct.SBand`. This permits the function to compute actual coverage probabilities, which is useful for evaluating the methodology. However, it is possible to supply a dummy vector here if the true function is not known.

A theoretical study of frequentist confidence intervals associated with wavelet shrinkage can be found in Picard and Tribouley (2000) and Genovese and Wasserman (2005).

4.7 Density Estimation

By now, there is a sizeable literature on wavelet density estimation. For example, see Kerkyacharian and Picard (1992a,b, 1993, 1997), Johnstone et al. (1992), Leblanc (1993, 1995, 1996), Masry (1994, 1997), Hall and Patil (1995, 1996), Hall and Wolff (1995), Delyon and Juditsky (1996), Donoho et al. (1996), Koo and Kim (1996), Huang (1997), Pinheiro and Vidakovic (1997), Safavi et al. (1997), Kato (1999), Muller and Vidakovic (1999), Zhang and Zheng (1999), Herrick (2000), Herrick et al. (2001), Hall and Penev (2001), Renaud (2002a,b), Juditsky and Lambert–Lacroix (2004), Ghorai and Yu (2004), Chicken and Cai (2005), Bezandry et al. (2005), and Walter and Shen (2005).

`WaveThresh` contains basic functionality for constructing wavelet based density estimators but very little of the sophisticated techniques necessary for complete density estimation. We follow the notation of Hall and Patil (1995). Suppose $f(x)$ is a density function (on the real line, or some interval) for some random variable X. Then we can expand $f(x)$ in terms of a wavelet expansion by

$$f(x) = \sum_k c_{0k}\phi_{0k}(x) + \sum_{j=0}^{J_{\max}} \sum_k d_{j,k}\psi_{j,k}(x), \qquad (4.16)$$

where the scaling function and wavelets have the following slightly modified definition from the standard one in Chapter 2:

$$\phi_{j,k}(x) = p_j^{1/2}\phi(p_j x - k),$$
$$\psi_{j,k}(x) = p_j^{1/2}\psi(p_j x - k),$$

for arbitrary $p > 0$ and $p_j = p2^j$. Here p is called the *primary resolution* and $J_{\max}$ the finest resolution level. The primary resolution here is related to the primary resolution described in Section 3.6, but it is not quite the same thing. In particular, our previous primary resolution was an integer, the one here is on a continuous scale, see Hall and Nason (1997) for a detailed description of the differences.

Since $f(x)$ is the probability density function for X note that

$$d_{j,k} = \int f(x)\psi_{j,k}(x)\, dx = \mathbb{E}\left\{\psi_{j,k}(X)\right\}, \tag{4.17}$$

and similarly $c_{0k} = \mathbb{E}\{\phi_{0k}(X)\}$. Hence, given an iid sample $X_1, \ldots, X_n$ from $f(x)$, the obvious empirical estimate of $d_{j,k}$ is obtained by replacing the population mean in (4.17) by the sample mean to give

$$\tilde{d}_{j,k} = n^{-1}\sum_{i=1}^{n}\psi_{j,k}(X_i), \tag{4.18}$$

and similarly to obtain an empirical estimate $\tilde{c}_{0k}$ of c_{0k}. Note that $\tilde{d}_{j,k}$ is an unbiased estimator of $d_{j,k}$. Fast methods of computation of these quantities are described in Herrick et al. (2001).

To obtain a nonlinear density estimate one then follows the usual wavelet paradigm and thresholds the empirical wavelet coefficients $\tilde{d}_{j,k}$ to $\hat{d}_{j,k}$ and then inverts the coefficients to obtain an estimate $\hat{f}(x)$. The function denproj in WaveThresh projects data $X_1, \ldots, X_n$ onto a scaling function basis at some resolution level j, i.e., it computes a formula similar to (4.18) but replacing ψ by ϕ. Then one can use the function denwd to compute the DWT of $X_1, \ldots, X_n$, which applies the fast pyramidal algorithm to the output of denproj. The functions denwr and denplot invert the transforms and plot the wavelet coefficients respectively. These functions make use of the Daubechies–Lagarias algorithm (Daubechies and Lagarias, 1992) to compute $\psi(x)$ efficiently.

The thresholding is not quite as straightforward as in the iid regression case earlier. For example, for a start, Herrick et al. (2001) show that the covariance of the empirical wavelet coefficients is given by

$$\mathrm{cov}\left(\tilde{d}_{j_1 k_1}\tilde{d}_{j_2 k_2}\right) = n^{-1}\left\{\int \psi_{j_1,k_1}(x)\psi_{j_2,k_2}(x)f(x)\, dx - d_{j_1,k_1}d_{j_2,k_2}\right\}. \tag{4.19}$$

Hence, the empirical wavelet coefficients are not iid as in the regression case. In particular, the variance of $\tilde{d}_{j,k}$ is given by

$$\mathrm{var}(\tilde{d}_{j,k}) = n^{-1}\left\{\int \psi_{j,k}^2(x)f(x)\, dx - d_{j,k}^2\right\}, \tag{4.20}$$

and this quantity can be calculated rapidly, if approximately, using the 'powers of wavelets' methods described in Section 4.6.1 (indeed, density estimation is where the idea originated). So wavelet coefficient variances for density estimation can be quite different from coefficient to coefficient.

Comparison to kernel density estimation. Let us consider the basic kernel density estimate of $f(x)$, see e.g., Silverman (1986), Wand and Jones (1994),

$$\breve{f}(x) = (nh)^{-1} \sum_{i=1}^{n} K\left(\frac{x - X_i}{h}\right), \tag{4.21}$$

where K is some kernel function with, $K(x) \geq 0$, K symmetric, and $\int K(x)\,dx = 1$.

The wavelet coefficients of the kernel density estimate, for some wavelet $\psi(x)$, are given by

$$\breve{d}_{j,k} = \int \breve{f}(x)\psi_{j,k}(x)\,dx$$

$$= (nh)^{-1} \sum_{i=1}^{n} \int K\left(\frac{x - X_i}{h}\right) \psi_{j,k}(x)dx$$

$$= n^{-1} \sum_{i=1}^{n} \int K(y)\psi_{j,k}(yh + X_i)dy, \tag{4.22}$$

after substituting $y = (x - X_i)/h$. Continuing

$$\breve{d}_{j,k} = \int K(y) n^{-1} \sum_{i=1}^{n} \psi_{j,k}(yh + X_i)\,dy$$

$$= \int K(y)\tilde{d}_{j,k-2^j yh}\,dy. \tag{4.23}$$

Hence, the wavelet coefficients of a kernel density estimate, $\breve{f}$, of f are just the kernel smooth of the empirical coefficients $\tilde{d}$. In practice, a kernel density estimate would not be calculated using Formula (4.23). However, it is instructive to compare Formula (4.23) to the nonlinear wavelet methods described above that threshold the $\tilde{d}$. Large/small local values of $\tilde{d}$ would, with good smoothing, still result in large/small values of $\breve{d}$, but they would be smoothed out rather than selected, as happens with thresholding.

Overall. The wavelet density estimator given in (4.16) is an example of an *orthogonal series estimator*, and like others in this class there is nothing to stop the estimator being negative, unlike the kernel estimator, which is always nonnegative (for a nonnegative kernel). On the other hand, a wavelet estimate might be more accurate for a 'sharp' density or one with discontinuities, and there are computational advantages in using wavelets. Hence, it would be useful to acquire positive wavelets. Unfortunately, for wavelets, a key property is $\int \psi(x)\,dx = 0$, and the only non-negative function to satisfy this is the zero function, which is not useful. However, it is possible to arrange for a wavelet-like function which is practically non-negative over a useful range of interest. Walter and Shen (2005) present such a construction called *Slepian semi-wavelets*, which only possess negligible negative values, retain the advantages of a wavelet-like construction, and appear to be most useful for density estimation.

Another approach to density estimation (with wavelets) is to bin the data and then apply appropriate (wavelet) regression methods to the binned data. This approach is described in the context of hazard density function estimation in the next section.

4.8 Survival Function Estimation

The problem of survival function estimation has been addressed in the literature using wavelet methods. Wavelets provide advantages in terms of computation speed but also for improved performance for survival functions that has sharp changes as often occurs in some real-life situations.

One of the earliest papers to consider wavelet hazard rate estimation in the presence of censoring was by Antoniadis et al. (1994), who proposed linear wavelet smoothing of the Nelson–Aalen estimator, see Ramlau–Hansen (1983), Aalen (1978). Patil (1997) is another early paper which considers wavelet hazard rate estimation with uncensored data.

For most of this section, we consider Antoniadis et al. (1999), where n subjects were considered with survival times $X_1, \ldots, X_n$ and (right) censoring times of $C_1, \ldots, C_n$. The observed random variables are Z_i and δ_i, where

$$Z_i = \min(X_i, C_i) \text{ and } \delta_i = \mathbb{I}(X_i \leq C_i),$$

where $\mathbb{I}$ is the indicator function. So, if $\delta_i = 1$, this means that $X_i \leq C_i$ and $Z_i = X_i$, the observed value is the true lifetime of subject i. If $\delta_i = 0$, this means that $X_i > C_i$ and hence $Z_i = C_i$ and so the actual lifetime X_i is not observed. A real example of this set-up might be studying cancer patients on a drug trial. If $\delta_i = 1$, then the observed variable Z_i is when the patient actually dies, whereas if $\delta_i = 0$, then true death time is not observed as something occurs for it not to be (e.g. the patient leaves the trial, or the trial is stopped early). Antoniadis et al. (1999) cite an example of times of unemployment. In this example, the 'lifetime' is the time from when a person loses their job until they find another one. This example is particularly interesting as there appear to be peaks in the estimate, not picked up by other methods, that appear to correspond to timespans when unemployment benefits cease.

Antoniadis et al. (1999) define $\{X_i\}_{i=1}^n$ and $\{C_i\}_{i=1}^n$ both to be nonnegative iid with common continuous cdfs of F and G respectively, and continuous density f and g respectively, and both sets are independent. The usual hazard rate function is given by $\lambda(t) = f(t)/\{1 - F(t)\}$, which expresses the risk of 'failing' in the interval $(t, t + \delta t)$ given that the individual has survived up to time t. Antoniadis et al. (1999) approach the problem as follows: if $G(t) < 1$, then the hazard rate can be written

$$\lambda(t) = \frac{f(t)\{1 - G(t)\}}{\{1 - F(t)\}\{1 - G(t)\}}, \tag{4.24}$$

for $F(t) < 1$. Then they define $L(t) = \mathbb{P}(Z_i \leq t)$, the observation distribution function, and since $1 - L(t) = \{1 - F(t)\}\{1 - G(t)\}$, and defining $f^*(t) = f(t)\{1 - G(t)\}$, the hazard function can be redefined as

$$\lambda(t) = \frac{f^*(t)}{1 - L(t)}, \tag{4.25}$$

with $L(t) < 1$. The quantity $f^*(t)$, termed the subdensity, has a density-like character. Antoniadis et al. (1999) choose to bin the observed failures into equally-spaced bins and use the proportion of observations falling in each bin as approximate estimators of $f^*(t)$. A better estimate can be obtained by a linear wavelet smoothing of the binned proportions. This binning/smoothing method is more related to the wavelet 'regression' methods described in Chapter 3 rather than the density estimation methods described in the previous section.

The $L(t)$ quantity in the denominator of (4.25) is estimated using the integral of a standard histogram estimator, which itself can be viewed as an integrated Haar wavelet transform of the data. See also Walter and Shen (2001, p. 301). The estimator of $\lambda(t)$ is obtained by dividing the estimator for $f^*(t)$ by that for $1 - L(t)$.

Although WaveThresh does not directly contain any code for computing survival or hazard rate function estimates, it is quite easy to generate code that implements estimators similar to that in Antoniadis et al. (1999). The construction of the subdensity $f^*(t)$ and $L(t)$ estimates can be written to make use of a 'binning' algorithm which makes use of existing R code: the table function as follows. First, a function bincount which, for each time Z_i, works out its bin location:

```
> bincount <- function(z, nbins=20){

+ zmin <- min(z)
+ zmax <- max(z)

+ prange <- zmax - zmin
+ prdel <- prange/nbins

#
# Extend the range a small amount so there is a bit of
# space at the beginning and end of the domain of
# definition
```

```
#
+ zmin <- zmin - prdel/2
+ zmax <- zmax + prdel/2

+ prange <- zmax - zmin
+ prdel <- prange/(nbins-1)

+ ans <- (z - zmin)/prdel
+ return(list(bincounts=ans,
+         bins=seq(from=zmin, to=zmax, by=prdel), del=prdel))
```

Here the bin width is quantified by the (final) `prdel` and the line that computes which bin each Z_i is in is `ans <- (z - zmin)/prdel`. The following function, `hfc`, counts how many Z_i are contained within each bin. Note, for simplicity, this implementation permits only dyadic bins.

```
> hfc <- function(z, nbins=32){

# Only permit dyadic num. of bins
+ if (is.na(IsPowerOfTwo(nbins)))
+         stop("nbins must be power of two")

# Will contain counts of each Zi
+ cc <- rep(0,nbins)

# Work out which bin each Zi belongs to
+ ans <- bincount(z, nbins=nbins)

# Count number of Zi in each bin
+ tb <- table(round(ans$bincounts))

+ ix <- as.numeric(dimnames(tb)[[1]])
+ cc[ix] <- tb

+ return(list(bincounts=cc, bins=ans$bins, del=ans$del))
+ }
```

Use of the `hfc` function merely counts how many Z_i occur in each bin. Since the scaling function of the Haar wavelet is a 'box function' as defined in Formula (2.22) these counts also happen to be the (estimated) father wavelet coefficients of the density at the resolution defined by the number of bins (parameter `prdel`). Indeed, these are exactly the density estimation coefficients as defined by (4.18), except $\tilde{d}_{j,k}$ and $\psi_{j,k}$ are replaced by $\tilde{c}_{j,k}$ and $\phi_{j,k}$ respectively. For Haar wavelets, the counts are a crude histogram estimate of the density of the Z_i. Antoniadis et al. (1999) advocate a simple linear smoothing of this 'fine-scale' histogram. Linear wavelet smoothing was

described in Section 3.11 and here can be achieved using the following function
hws:

```
> hws <- function (yy, lev=1)
+ {
# Do Haar WT of fine-scale counts
+ yywd <- wd(yy, filter.number=1, family="DaubExPhase")

# Zero out a whole set of levels of coefficients
+ levs <- nlevels(yywd) - (1:lev)
+ yywdT <- nullevels(yywd, levelstonull=levs)

# Return the reconstruction of the remainder
+ wr(yywdT)
}
```

By default hws sets the finest-scale coefficients to zero, but successively coarser
levels can be zeroed by increasing the lev parameter.

The estimator of $L(t)$ is constructed by integrating the output of the linear
wavelet smoothing, or rather an approximation of it, the cumulative sum as
follows:

```
> Lest <- function(z, nbins=32, lev=1){

+ ans <- hfc(z, nbins=nbins)
+ L <- 1-cumsum(hws(ans$bincounts, lev=lev))/length(z)

+ ans <- list(L=L, bins=ans$bins, del=ans$del)
+ ans
}
```

In fact, Lest, the estimator of $L(t)$, is a summed histogram, as specified in
Antoniadis et al. (1999), and is a form of (cumulative) density estimator,
albeit a very simple one. For example, one can produce a survival function
plot by typing

```
# Invent some data
> zz <- rnorm(200)

> nbins <- 32
> barplot(Lest(zz, nbins=nbins, lev=1)$L, ylim=c(0,1))
```

This just 'smooths' by zeroing the finest-level coefficients. Plots of estimates that zero out the finest one, two, three, and four levels are shown
in Figure 4.15.

The subdensity estimate can be computed in a similar way. Code to
compute an estimate of $f^*(t)$ can be found in Appendix C along with a routine
to compute the hazard function estimate itself.

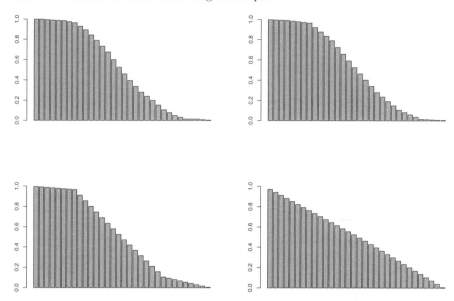

Fig. 4.15. Linear wavelet SF estimators produced by Lest function on sample Gaussian data. *Clockwise* from *top-left*: zeroing 1, 2, 4 and 3 finer resolution levels.

Antoniadis et al. (1999) go on to demonstrate pointwise and global mean-square consistency, best possible asymptotic MISE convergence rates (under mild smoothness conditions), and asymptotic normality of the estimator. Bezandry et al. (2005) recently proved (weak) *uniform* consistency of the subdensity and hazard rate estimators, which suggests that good point estimates, along with confidence intervals, can be constructed from samples of a reasonable size.

Antoniadis et al. (2000) describe a method for estimating the location of a change point in an otherwise smooth hazard function with censored data. MISE results for hazard rate estimation for censored survival data via a wavelet–Kaplan–Meier approach can be found in Li (2002). Nonlinear hazard rate estimation in the presence of censoring and truncation was considered by Wu and Wells (2003), who proposed direct nonlinear wavelet shrinkage of the Nelson–Aalen cumulative hazard estimator, i.e., wavelet shrinkage of the hazard itself rather than the two-step, $f^*(t), L(t)$ procedure above. Rodríguez–Casal and De Uña–Álvarez (2004) consider nonlinear wavelet density estimation under the assumption of the Koziol–Green random censorship model where the censoring survival function is a power of the lifetime survival function, i.e., $1 - G(t) = \{1 - F(t)\}^\beta$, for some parameter β. Bouman et al. (2005) describe a 'wavelet-like' Bayesian multiresolution model and estimator based on a binary tree construction. Liang et al. (2005) examine the global L_2 error of nonlinear estimators of the density and hazard rate estimators and establish optimal convergence rates for such functions in Besov classes.

4.9 Inverse Problems

To begin we follow the development and notation in Abramovich and Silverman (1998). Unlike the *direct* problems above, suppose one wishes to learn about an unknown function, $f(t)$, but one only has access to $(Lf)(t)$, where L is some linear operator. For example, L might be the integration operator. This is a linear inverse problem. Further, suppose that the data are observed discretely in the presence of iid Gaussian noise, $\{\epsilon_i\}$, i.e., we observe

$$y_i = (Lf)(t_i) + \epsilon_i. \qquad (4.26)$$

Inferring f from y is known as a statistical linear inverse problem, and O'Sullivan (1986) is a principal reference. Often, the problem is *ill-posed*, which means that it is not straightforward to obtain f from Lf (even without the noise) in that a solution might not exist or solutions might not be unique.

There are several approaches to tackling inverse problems. A convenient approach is the *method of regularization*, due to Tikhonov (1963), which essentially tries to choose the best f that fits the data y through (4.26) but additionally applies some kind of smoothness or sparsity, or, depending on the problem, some other condition of 'reasonability' on the solution.

Another method is to use the singular value decomposition (SVD) paradigm in which the unknown function f is expanded in a series of eigenfunctions that depend on the operator. It can be shown (Johnstone and Silverman, 1990, 1991), that the asymptotically best estimator, in a certain minimax sense, can be obtained by a properly truncated expansion in the eigenbasis, over functions f, that display homogeneous variation. However, as Abramovich and Silverman (1998) point out, there is sometimes a mismatch between the smoothness of the eigenbasis used to estimate f and the class of functions that you might wish f to belong to. For example, if the operator L is such that the eigenbasis is the Fourier basis (stationary operators), then these might not be suitable for representing interesting functions f which might contain discontinuities, for example, edges in an image.

Partly in response to this mismatch Donoho (1995b) proposed the *wavelet-vaguelette decomposition* (WVD), where the unknown function f is expanded in terms of a wavelet expansion, then an associated 'vaguelette' series for Lf is constructed, and finally coefficients are estimated using thresholding (and these coefficients then form the coefficients for the wavelet representation of f). The key point underlying this method is that, given a suitable family of wavelets $\{\psi_{j,k}\}$, for some special operators, L, there exist constants, $\beta_{j,k}$, such that the set of scaled function $v_{j,k} = \beta_{j,k} L\psi_{j,k}$ forms a (Riesz) basis. Kolaczyk (1994, 1996) explains in detail how the WVD can be used for tomographic image reconstruction; here the linear operator L is the Radon transform: the integral of a 2D function over a line.

Abramovich and Silverman (1998) propose the *vaguelette-wavelet decomposition* (VWD), which expands Lf in terms of wavelets, and the estimated

f is expressed in terms of the associated vaguelette expansion. The thresholding is conceptually more attractive in the VWD case as it is the standard thresholding of iid data as described in Chapter 3. In some practical examples based on the inhomogenous test functions from Section 3.4, the WVD/VWD methods performed similarly to and better than those of a SVD approach.

Johnstone et al. (2004) study the deconvolution inverse problem in some depth and consider, particularly, boxcar convolution. Here, the L operator is the convolution operator, i.e. $h(t) = (Lf)(t) = (f * g)(t) = \int_T f(t - u)g(u)\,du$, where $g(t)$ is some blurring function. Their work necessarily involves the Fourier transform as Fourier converts convolutions into multiplications. In other words, if h_l, f_l, g_l are the Fourier coefficients of $h(x), f(x), g(x)$ respectively, then $h_l = f_l \times g_l$. Hence, f_l can be recovered from h_l/g_l, and so the properties of g_l in the denominator are crucial.

A *boxcar* blur is simply the indicator function of some interval: Johnstone et al. (2004) define the boxcar function $g(x)$ to be $g(x) = (1/2a)\mathbb{I}_{[-a,a]}(x)$, where a is some spatial scale, and note that the Fourier coefficients of g are

$$g_l = \frac{\sin(\pi la)}{\pi la}, \tag{4.27}$$

for $l \in \mathbb{Z}$. Hence, problems of a zero denominator in h_l/g_l occur whenever $g_l = 0$, which occurs when $a = p/q$ is a rational number and l is any integer multiple of q. Hence, recovery cannot happen for these frequencies. The problem is not so severe for irrational numbers and particularly those that are far from ('badly approximable' by) rationals.

The Johnstone et al. (2004) *WaveD* method is particularly elegant. They note that the wavelet coefficients of the unknown function $f(x)$ can be written using Plancheral's equality as

$$\beta_\kappa = \int f(x)\Psi_\kappa(x)\,dx = \sum_l f_l \Psi_l^\kappa, \tag{4.28}$$

where $\Psi_\kappa(x)$ is the κth (periodized) wavelet and $\{\Psi_l^\kappa\}_l$ are the Fourier coefficients of $\Psi_\kappa(x)$. Ideally, one could then replace f_l by h_l/g_l in (4.28), but from (4.26) we observe y and not $h = Lf$. So Johnstone et al. (2004) take y_l (the Fourier coefficients of what is observed) as an unbiased estimator of h_l, and their estimator for the wavelet coefficients of f is given by

$$\hat{\beta}_\kappa = \sum_l \frac{y_l}{g_l}\Psi_l^\kappa, \tag{4.29}$$

from which f can be recovered using an inverse wavelet transform. One can see that f is expanded in a wavelet basis and, as Johnstone et al. (2004) note, their estimator "*formally* can be viewed as being consistent with the WVD recipe". However, their implementation is different in that it operates in the Fourier domain with y_l, g_l and the like. Johnstone et al. (2004) compare their WaveD with a linear Fourier regularized deconvolution method and a wavelet-regularized deconvolution method due to Neelamani et al. (2004), who also coin the terms FoRD and ForWaRD for these methods. Johnstone et al. (2004) note that the regularized Fourier filtering distorts the original signal and WaveD is better for higher noise levels. For smooth blur WaveD beats both FoRD and ForWaRD. For boxcar blur WaveD outperforms ForWaRD for high noise levels, but the situation is reversed for smaller noise levels. Both WaveD and ForWaRD outperform FoRD in this situation as the latter is linear. Further details on WaveD can be found in e.g., Donoho and Raimondo (2004) on mathematical details and e.g., Cavalier and Raimondo (2007), which considers the case where g itself needs to be estimated and is noisy. Cavalier and Raimondo (2007) also provide a useful review of the growing body of literature in this area.

`WaveThresh` does not (directly) provide any facilities for the above inverse problem techniques. However, Raimondo and Stewart (2007) have served the community admirably by implementing WaveD for R as the `waved` package available from CRAN.

5

Multiscale Time Series Analysis

5.1 Introduction

The modelling and analysis of dependent phenomena are among the most important and challenging areas of statistics. Put simply, a time series is a set of observations with some definite ordering in time. For example, we may denote an observed (discrete) time series of length n by $x_1, \ldots, x_n$. The main difference between time series data and 'ordinary' data is that time series observations are not independent but typically possess a stochastic relationship *between* observations (here we are already assuming a stochastic model for our data).

A great deal has been written about time series analysis, and we will not repeat it here. We assume that the reader is familiar with time series at the level of the introductory book by Chatfield (2003), although we will reintroduce some basic concepts where necessary. There are many other equally excellent books on the topic such as Hannan (1960), Priestley (1983), Brockwell and Davis (1991), Hamilton (1994), West and Harrison (1997), and Brillinger (2001).

Time series arise in all sorts of areas—it is difficult to name one where they do not. As for statistics generally, the *statistical* analysis of time series usually consists of developing a model for the series, fitting the model, assessing the model, and then, in response to the assessment, reformulating the model and repeating the procedure as necessary. The aim of such an analysis might be to *model* the series, possibly to obtain some parsimonious mathematical description of the mechanics underlying the data generation. Often, another important goal is to *forecast* future values of the series based on data collected up to a given time point. With multiple time series one might well be interested in modelling and forecasting but also studying the relationship between the series and, for example, using the values of one series to predict another.

Many methods already exist to carry out these tasks. This chapter investigates wavelet methods which can help with these aims. For time series analysis sparsity of wavelet representations is probably less important, but the wavelet ability for localization is more important. For stationary series of particular types, the wavelet ability to concentrate information in certain dyadic scales is a useful property. For locally stationary series, the ability of wavelets to localize information in a systematic manner makes them a good choice for the local time-scale analysis of information within a series. However, above all, it is important to remember that wavelets are really just tools, like other tools such as Fourier. Wavelets might be a useful tool in the modelling and analysis of a given time series, but it might be that a non-wavelet tool performs better. These kinds of modelling questions are still open for investigation.

The most general probability model for a time series consists of the specification of the joint probability distribution of the collection of random variables $(X_{t_1}, \ldots, X_{t_n})$ for all $(t_1, \ldots, t_n)$ and integers $n \geq 1$. However, this is almost always too general and simplifications are almost always made.

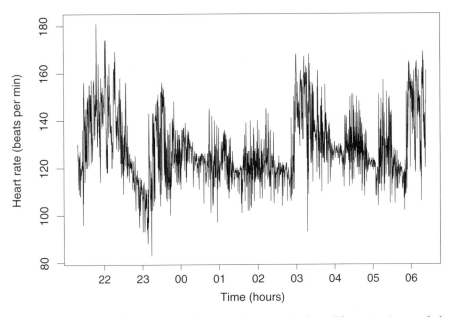

Fig. 5.1. Electrocardiogram recording of a 66-day-old infant. The series is sampled at 1/16Hz and is recorded from 21:17:59 to 06:27:18. There are 2048 observations. Reproduced from Nason et al. (2000) with permission.

This chapter will illustrate multiscale techniques using two particular time series. The first series, collected by the Institute of Child Health at the University of Bristol, is a medical data set that records the heart rate of an infant during a night. The data are available within WaveThresh after issuing a data(BabyECG) command. The data set is displayed in Figure 5.1.

The main point to note about the BabyECG time series is that it is unlikely to be stationary (see below for a formal definition of stationarity). The pattern of the ECG varies markedly across the night and experiences transitions from one regime to the next as the baby enters different periods of sleep and shifts in physical position. If the reader still thinks that the `BabyECG` time series is stationary, then they are invited to compute the empirical autocorrelation function using the `acf()` function on two non-overlapping portions of the series: you will almost certainly get different answers.

Despite the non-stationarity, it is still interesting to contemplate an analysis of the ECG series but by assuming it is a locally stationary series (which, again, we will formally define below). The `BabyECG` series also seems to possess fairly sharp transitions between different regimes: compactly supported wavelets are well suited to this kind of phenomenon. The other point to note with the `BabyECG` series is that it comes as a pair with another time series: the sleep state, which can be loaded in from `WaveThresh` using `data(BabySS)`. The sleep state series is a discrete-valued time series taking the values 1='quiet sleep', 2='between states 1 and 3', 3='active sleep', and 4=awake. One aim for the analysis of these time series was to ask whether it was possible to predict sleep state successfully from the ECG alone. The reason being that sleep state itself is very expensive and time consuming to determine from videos of the infants, whereas ECG recording is quick and easy and can be obtained routinely within an infant's natural home environment.

Our second time series example concerns the Realtime Seismic Amplitude Measurement (RSAM) counts taken on Montserrat shown in Figure 5.2. These counts are 10-minute averages, sampled every 5 minutes, of ground shaking caused by earthquakes and volcanic tremor. Again, it is unlikely that the series is stationary. For one thing, the variation of the series appears to be greater in the second half of the series compared to the first half. The other aspect of this series is that the variance in the series might be linked to the mean: a common feature of count data.

5.2 Stationary Time Series

5.2.1 Stationary models and cartoons

This section recalls the notion of stationary time series and show the benefits for computer simulation in understanding properties of time series estimation procedures over and above those provided by a theoretical analysis. It also allows us to introduce some concepts and notation that will be of use later in the chapter.

The statistical literature on time series analysis is dominated by the theory and practice of *stationary* series. Informally, a stationary time series is one whose statistical properties do not change over time. More formally, a *strictly stationary* time series is one where the joint distribution of $(X_{t_1}, \ldots, X_{t_n})$ is the same as $(X_{t_1+\tau}, \ldots, X_{t_n+\tau})$ for all t_i, n, and τ.

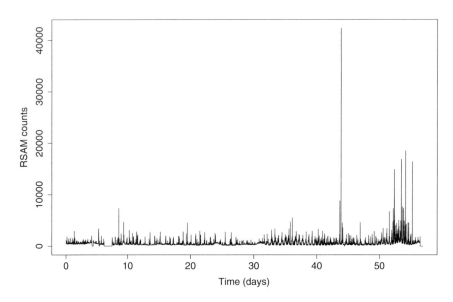

Fig. 5.2. RSAM count series from Montserrat from the short-period station MBLG, Long Ground, 16:43.50N, 62:09.74W, 287m altitude. Series starts 31 July 2001 00:00, ends 8 October 2001 23:54. There are 16384 observations. Reproduced from Nason (2006) with permission from the International Association of Volcanology and Chemistry of the Earth's Interior.

However, strict stationarity is usually too strong for most practical purposes, and the following weaker form is used. A time series is said to be *second-order* or *weakly stationary* if the following holds $\mathbb{E}(X_t) = \mu$, and the autocovariance $\gamma(\tau) = \text{cov}(X_t, X_{t+\tau})$ is a function of τ only (the latter also implies that the variance of X_t is constant as a function of time, all t, τ being integers). The process autocovariance $\gamma(\tau)$ can be estimated by the *sample autocovariance*, $c(\tau)$, which is obtained by calculating the usual sample covariance on the sample values $\{x_t\}_{t=1}^{T-\tau}$ with lagged values $\{x_{t+\tau}\}_{t=1}^{T-\tau}$. The autocorrelation measures the extent of linear association between x_t and lagged values $x_{t+\tau}$ and hence measures the degree of 'internal linear relationships' within the time series at different lags. The *partial autocorrelation* at lag τ is the excess autocorrelation at lag τ that is not already accounted for by autocorrelation at lower lags.

The spectrum, or spectral density function, $f(\omega)$, is a measure of the 'amount' of oscillation at different frequencies $\omega \in (-\pi, \pi)$. Specifically, $f(\omega)d\omega$ is the contribution to the total variance of X_t for frequencies in the range $(\omega, \omega + d\omega)$. The spectrum is related to the autocovariance via the following Fourier relationship $f(\omega) = (2\pi)^{-1} \sum_{\tau=-\infty}^{\infty} \gamma(\tau) \exp(-i\omega\tau)$.

A *purely random* time series model or process, $\{Z_t\}_{t \in \mathbb{Z}}$, is a collection of independent and identically distributed random variables. For example, they could be $Z_t \sim N(0, \sigma^2)$.

Purely to exhibit our simulations we shall use the following well-known time series model class. If X_t is an *autoregressive moving average* (ARMA) model of order (p, q), then X_t has a representation given by

$$X_t = \sum_{i=1}^{p} \alpha_i X_{t-i} + Z_t + \sum_{j=1}^{q} \beta_j Z_{t-j}. \tag{5.1}$$

This is all for positive integers p, q and (here real-valued) model parameters $\{\alpha_i\}_{i=1}^{p}$ and $\{\beta_j\}_{j=1}^{q}$.

Within R it is extremely easy to simulate an ARMA process. For example, suppose we wished to simulate 1000 observations from the ARMA(2,1) process with AR parameters $(\alpha_1, \alpha_2) = (0.8, -0.5)$ and MA parameter of 1.5. Then we could issue the command

```
> x <- arima.sim(1000, model=list(ar=c(0.8, -0.5), ma=1.5,
+       order=c(2,0,1)))

> ts.plot(x)
```

Such a realization is depicted in the top left-hand plot of Figure 5.3. We can also use the regular R functions for computing the autocorrelation, partial autocorrelation, and spectral density function as depicted in the other three subplots of Figure 5.3.

Of course, the autocorrelation and spectral quantities shown in Figure 5.3 are subject to sampling variation. Another realization, say another run of function f.tsa1(), would result in different autocorrelation and spectrum functions. The beauty of statistical computation means that we can obtain an accurate idea of what the true autocorrelation and spectral quantities are for the underlying ARMA process by repeatedly simulating from the same model, computing the quantities, and averaging the result. The results of this averaging process for 100 simulations are shown in Figure 5.4. In this plot, good, or cartoon, estimates of all the quantities are shown, e.g. notice the near-zero mean.

Although these kinds of simulation techniques are not necessarily useful for data analyses, they are useful for understanding the properties of time series analysis methods—especially when it is difficult to analyze methods by theoretical means. Such simulations are extremely useful when studying locally stationary processes, as we shall see later in this chapter.

5.2.2 Scale analysis of stationary time series

Wavelet methods can be useful for stationary time series. There exist analogues of the process variance, autocorrelation, cross-correlation, and spectrum but adapted and indexed by scale. Many of these methods rely on the

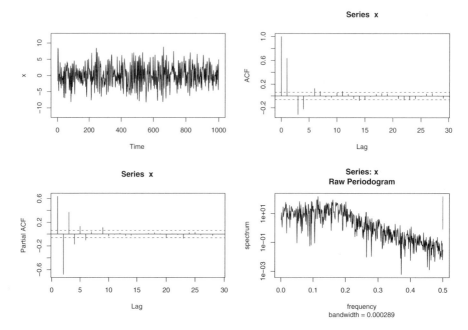

Fig. 5.3. *Top left*: realization of a single simulation of 1000 observations of an ARMA(2,1) process with AR parameters $(\alpha_1, \alpha_2) = (0.8, -0.5)$ and MA parameter of 1.5. *Top right*: empirical autocorrelation of realization. *Bottom left*: empirical partial autocorrelation of realization. *Bottom right*: spectral estimate of realization using the `spectrum()` function of R. Produced by `f.tsa1()`.

classical discrete wavelet transform, better estimates tend to be produced by non-decimated versions, and several methods use more complicated techniques such as non-decimated wavelet packets. A good example is Chiann and Morettin (1999), who introduce a wavelet spectral analysis for a stationary discrete process including a wavelet version of the periodogram and consideration of the scalegram (the energy of the wavelet coefficients at a particular scale). For an introduction and comprehensive review of the field see Percival and Walden (2000) or Gencay et al. (2001).

For some methods, though, one can question the reason behind using multiscale methods. For example, some formulations of the wavelet spectrum of a stationary time series are merely a simple integration of the regular Fourier spectrum over dyadic frequency bands. So, generally, a wavelet spectrum can be less informative than the Fourier spectrum, and it is sometimes difficult to see what benefit the wavelet methods bring. On the other hand, if one has prior information that the relevant information in a time series is somehow scale, and only scale, related, then it is likely that a wavelet spectral method is an appropriate tool.

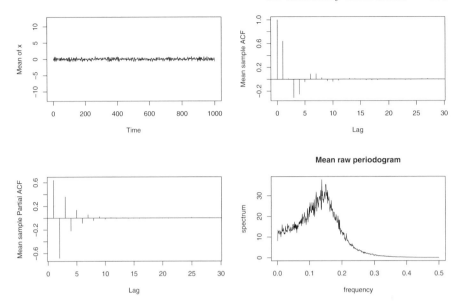

Fig. 5.4. *Top left*: mean of 100 realizations an ARMA(2,0,1) process with AR parameters $(\alpha_1, \alpha_2) = (0.8, -0.5)$ and MA parameter of 1.5. *Top right*: average empirical autocorrelation over realizations. *Bottom left*: average empirical partial autocorrelation. *Bottom right*: average spectral estimate. Produced by `f.tsa2()`.

For example, Percival and Guttorp (1994) and Percival (1995) introduced the *wavelet variance*, which is a means of discovering how much variance exists within a process at different dyadic scales. A simple example of the wavelet variance is the Allan variance, which is a measure of stability for clocks and oscillators, see Allan (1966). Following the exposition in Nason and von Sachs (1999), the *Allan variance*, $\sigma_X^2(\tau)$, at a particular scale $\tau \in \mathbb{Z}$ measures how averages of X_t, over windows of length τ, change over time. Defining $\bar{X}_t(\tau) = \tau^{-1} \sum_{n=0}^{\tau-1} X_{t-n}$, then

$$\sigma_X^2(\tau) = \frac{1}{2} \mathbb{E}\left(\left| \bar{X}_t(\tau) - \bar{X}_{t-\tau}(\tau) \right|^2 \right). \tag{5.2}$$

Here, we assume stationarity of X_t (although for a sensible estimator X_t can be a little more general). It turns out that $\sigma_X^2(\tau)$ can be written in terms of Haar wavelets. For example,

$$\sigma_X^2(1) = \frac{1}{2} \mathbb{E}\left(|X_t - X_{t-1}|^2 \right)$$

$$= \frac{1}{2} \mathbb{E}\left(|X_{2k+1} - X_{2k}|^2 \right)$$

$$= \frac{1}{2} \mathbb{E}\left(\left| \sqrt{2} d_{1,k} \right|^2 \right) = \mathbb{E}\left(d_{1,k}^2 \right) = \mathrm{var}(d_{1,k}),$$

where $d_{1,k}$ are the finest-scale Haar wavelet coefficients of X_t (and again assuming stationarity). Indeed, Nason and von Sachs (1999) note that by letting $\tau_j = 2^{-j-1}$ for $j \leq 1$ (where $j = -1$ is the finest scale and $j < -1$ corresponds to increasingly coarser scales) the (dyadic) Allan variance can be written as

$$\sigma_X^2(\tau_j) = \tau_j^{-1}\mathrm{var}(d_{j,k}), \tag{5.3}$$

and this suggests that the discrete wavelet transform coefficients could form the basis of an estimator for the Allan variance at dyadic scales. The wavelet variance is similarly defined but using more general wavelets. Percival (1995) considers estimators based on the decimated wavelet coefficients (Section 2.7) and on non-decimated wavelet coefficients (Section 2.9) and shows that the latter have a more favourable asymptotic relative efficiency. The reason for this is not surprising. For example, with the Haar wavelet variance the non-decimated estimator includes information on $X_3 - X_2$, for example, whereas the decimated version does not (i.e. decimation throws away much information).

The reader might wonder about the use of quantities such as the wavelet variance especially since a Fourier spectral estimate appears to be a richer source of information (and indeed the Fourier representation of a stationary time series is canonical). However, there are classes of processes for which the wavelet variance is appropriate. For example, Percival (1995) notes that the important 'power law processes', ones with spectrum $f(\omega) \propto |\omega|^\alpha$ over a range of frequencies, have wavelet variances where $\sigma^2(\tau) \propto \tau^{-\alpha-1}$ over a corresponding set of scales. Hence, a log-log plot of the wavelet variance against scale can help identify the power law parameter α. Another way of looking at this is to see that the wavelet, or Allan, variance provides a sparse representation of the available information. For such processes, we do not need the full flexibility of a spectral estimate, and the information can sufficiently be represented by the wavelet variance.

More recently Gabbanini et al. (2004) extended the definition of wavelet variance to wavelet packets (Section 2.11) and introduced a method to discover variance change points based on wavelet packet variance.

5.3 Locally Stationary Time Series

This section considers the recently introduced class of locally stationary stochastic processes known as *locally stationary wavelet (LSW) processes* introduced by Nason et al. (2000). It is necessary to point out that LSW processes join a growing list of useful process models stemming from Dahlhaus (1997) but with origins in such models as the *oscillatory* model introduced by Priestley (1965) and others such as Silverman (1957). We motivate the development of LSW processes by recalling the theoretical models for stationary series.

If a time series $\{X_t\}_{t \in \mathbb{Z}}$ is a stationary stochastic process, then it admits the following representation, see Priestley (1983):

$$X_t = \int_{-\pi}^{\pi} A(\omega) \exp(i\omega t) d\xi(\omega), \qquad (5.4)$$

where $A(\omega)$ is the amplitude of the process and $d\xi(\omega)$ is an orthonormal increments process. Put simply the process X_t is the sum (integration) of a collection of sinusoids at different frequencies, $\exp(i\omega t)$, for $\omega \in (-\pi, \pi)$ where the oscillation at frequency ω is 'magnified' by $A(\omega)$. So, if $A(\omega)$ is large for a given frequency, ω^*, say, and much larger than other $A(\omega)$, then a (stochastic) oscillation of frequency ω^* appears as a dominant feature in a realization of the process.

The point with stationary processes, and the representation in (5.4), is that the amplitude $A(\omega)$ does not depend on time. That is, the amplitude function is the same for the process over all time. Hence, the frequency behaviour of the time series is the same for all time. For many real time series, such as those depicted in Figures 5.1 and 5.2, this is just not the case, and model (5.4) is not adequate in any sense. One way of introducing time dependence into a representation such as (5.4) is to replace $A(\omega)$ by a time-dependent form such as $A_t(\omega)$. This is the kind of idea promoted by Priestley (1965) and Dahlhaus (1997), which results in a time-*frequency* model.

Given an actual time series, how might one discover whether a locally stationary model is appropriate? One way is to test to see if the local spectrum (however defined) varies over time. See, for example, Priestley and Subba Rao (1969) or von Sachs and Neumann (2000).

Another, multiscale, approach to locally stationary time series was introduced by Nason et al. (2000), who created a rigorous time-*scale* model by replacing the set of Fourier functions $\{\exp(i\omega t)\}, \omega \in (-\pi, \pi)\}$, by a set of discrete non-decimated wavelets. The wavelet-based model is of value for at least three reasons:

1. In the same way that wavelet variance efficiently described 'power law' processes in Section 5.2.2, locally stationary processes based on wavelets are effective for the modelling and analysis of time series that exhibit time-*scale* spectral variation.
2. Many developments for non-stationarity involve modifications to the Fourier representation in (5.4). Introduction of a wavelet-based model emphasizes that, for nonstationary processes, Fourier is not necessary and that other basis functions might well be of use.
3. The wavelet-based model we develop is pedagogically attractive as it does not require an understanding of stochastic integration.

The building blocks of our locally stationary wavelet model are described next.

5.3.1 Discrete non-decimated wavelets

This section defines our basic building blocks for constructing our discrete-time LSW processes. Let $\{h_k\}$ and $\{g_k\}$ be the low- and high-pass quadrature mirror filters that are used in the construction of Daubechies' compactly supported wavelets as in Section 2.3.3. Nason et al. (2000) constructed the compactly supported *discrete wavelets* $\psi_j = (\psi_{j,0}, \ldots, \psi_{j,(N_j-1)})$ of length N_j for scale $j < 0$ using the following formulae:

$$\psi_{-1,n} = \sum_k g_{n-2k}\delta_{0,k} = g_n, \text{ for } n = 0, \ldots, N_{-1} - 1, \tag{5.5}$$

$$\psi_{(j-1),n} = \sum_k h_{n-2k}\psi_{j,k}, \text{ for } n = 0, \ldots, N_{j-1} - 1, \tag{5.6}$$

$$N_j = (2^{-j} - 1)(N_h - 1) + 1, \tag{5.7}$$

where $\delta_{0,k}$ is the Kronecker delta, (B.4), and N_h is the number of non-zero elements of $\{h_k\}$. These formulae can be derived from the inverse DWT described in Section 2.7.4, specifically Formula (2.93).

For example, the discrete Haar wavelets at scales -1 and -2 respectively are

$$\psi_{-1} = (g_0, g_1) = (1, -1)/\sqrt{2}, \tag{5.8}$$
$$\psi_{-2} = (h_0 g_0, h_1 g_0, h_0 g_1, h_1 g_1) = (1, 1, -1, -1)/2, \tag{5.9}$$

and so on. As Nason et al. (2000) point out, the discrete wavelets are precisely the vectors produced by Daubechies' cascade algorithm used for producing discrete approximations to continuous-time wavelets at successively finer scales.

The discrete wavelets can be produced very easily using basic commands in WaveThresh. Many WaveThresh commands we use here were defined in Chapter 2, particularly Sections 2.9 and 2.11 For example, to produce the discrete Haar wavelets above, first set up a zero vector and transform it with the discrete wavelet transform:

```
> zwd <- wd(rep(0,16), filter.number=1, family="DaubExPhase")
```

Then put in a one at the finest resolution level so that on reconstruction only one step of the inverse DWT is applied (since all other coefficients are zero):

```
> wr(putD(zwd, level=3, c(0,0,0,0,1,0,0,0)))
 [1]  0.0000000  0.0000000  0.0000000  0.0000000  0.0000000
 [6]  0.0000000  0.0000000  0.0000000  0.7071068 -0.7071068
[11]  0.0000000  0.0000000  0.0000000  0.0000000  0.0000000
[16]  0.0000000
```

and the vector $(1, -1)/\sqrt{2}$ can be discerned: the code implements Equation (5.8) and the 1 is the δ function. For the discrete wavelet at scale -2 one can put in a one at the next coarsest scale as follows:

```
> wr(putD(zwd, level=2, c(0,0,1,0)))
 [1]  0.0  0.0  0.0  0.0  0.0  0.0  0.0  0.0
 [9]  0.5  0.5 -0.5 -0.5  0.0  0.0  0.0  0.0
```

and again one can find $\psi_{-2} = (1,1,-1,-1)/2$. It is very easy to repeat the above for other Daubechies' wavelets. For example, with the extremal-phase wavelet with two vanishing moments we first create the appropriate zero vector, transform it with the appropriate DWT, and then reconstruct using the δ function:

```
> zwd <- wd(rep(0,16), filter.number=2, family="DaubExPhase")
wr(putD(zwd, level=3, c(0,0,0,0,1,0,0,0)))
 [1]  0.0000000  0.0000000  0.0000000  0.0000000  0.0000000
 [6]  0.0000000 -0.1294095 -0.2241439  0.8365163 -0.4829629
[11]  0.0000000  0.0000000  0.0000000  0.0000000  0.0000000
[16]  0.0000000
```

Hence, for this Daubechies' wavelet $\psi_1 \approx (-0.1294, -0.2241, 0.8365, -0.4830)$.

Non-decimated discrete wavelets permit a wavelet to appear at each time point at each scale so that $\psi_{j,k}(\tau) = \psi_{j,(k-\tau)}$. This is in contrast to decimated wavelets, where their positioning depends on their scale, i.e., coarse-scale wavelets are more widely spaced than fine-scale wavelets.

5.3.2 Locally stationary wavelet (LSW) processes

Now that we have defined our oscillatory building blocks (the non-decimated discrete wavelets) we are in a position to specify our time series model based on them.

A *LSW process* $\{X_{t,T}\}_{t=0,...,T-1}$, $T = 2^J \geq 1$ is a doubly-indexed stochastic process having the following representation in the mean-square sense:

$$X_{t,T} = \sum_{j=-J}^{-1} \sum_k w_{j,k;T} \psi_{j,k}(t) \xi_{j,k}, \tag{5.10}$$

where $\{\xi_{j,k}\}$ is a random orthonormal increment sequence, $\{\psi_{j,k}(t) = \psi_{j,k-t}\}_{j,k}$ is a set of discrete non-decimated wavelets, and $\{w_{j,k;T}\}$ is a set of amplitudes. Although the representation in (5.10) may look complicated, it is actually very simple. The representation merely says that we are building a time series model $X_{t,T}$ out of a linear combination of oscillatory functions ($\psi_{j,k}$) with random amplitudes ($w_{j,k;T}\xi_{j,k}$) which is the multiscale equivalent of the building process for the stationary processes in (5.4).

Nason et al. (2000) specify three sets of conditions on the quantities in representation (5.10). The first is that $\mathbb{E}(\xi_{j,k}) = 0$, so that the process $X_{t,T}$ is always zero mean. In practice, any non-zero mean of the time series can be estimated and removed. The second condition is that the orthonormal increment sequence is uncorrelated. That is, $\text{cov}(\xi_{j,k}, \xi_{\ell,m}) = \delta_{j,\ell}\delta_{k,m}$, where $\delta_{j,\ell}$ is the Kronecker delta, (B.4).

It might not be immediately obvious from looking at the LSW representation in (5.10) that it actually *has* time-dependent properties. At first glance, the amplitudes, $w_{j,k;T}$, of the process do not depend directly on time. However, they do depend on k, and for a given time point t the discrete wavelet $\psi_{j,k}(t)$ has compact support localized around t and so only 'allows in' some $w_{j,k;T}$ near to t. Another set of $w_{j,k;T}$ is 'allowed in' for another t. So the time dependence of the statistical properties of $X_{t,T}$ relies on the amplitudes $w_{j,k;T}$ but indirectly through the localized discrete wavelet.

Hence, the statistical properties of $X_{t,T}$ do depend on the speed of evolution of $w_{j,k;T}$ as a function of k, and this happens for each scale j. The third condition from Nason et al. (2000) controls the speed of evolution of $w_{j,k;T}$ by forbidding it to deviate too much from a function $W_j(z)$ for $z \in (0,1)$ by assuming

$$\sup_k |w_{j,k;T} - W_j(k/T)| \leq C_j/T, \tag{5.11}$$

where $\{C_j\}$ is a set of constants with $\sum_{j=-\infty}^{-1} C_j < \infty$. Smoothness constraints are imposed on $W_j(z)$, which prevents it from oscillating too wildly, and this controls the speed of evolution of $w_{j,k;T}$. For a good detailed discussion on possible (and correct!) constraint sets on the model parameters, especially $W_j(z)$, see Fryzlewicz (2003).

The reason for not wanting the statistical properties of $X_{t,T}$ to evolve too rapidly is to permit estimation. The slower a process evolves, the larger the set of observations that can be pooled to obtain good estimates of the process generator W_j (or, more properly, its square, as described below). The following example from Nason and von Sachs (1999) indicates the possible estimation problems with series that evolve too fast. If we have a series X_t where $\sigma_t^2 = \text{var}(X_t)$ with different values of σ_t^2 for each t, then we have only one time series observation to estimate each σ_t^2. However, if σ_t changed very slowly, then we could use X_t and pool neighbouring values to get better estimates.

Example 5.1 (Haar MA processes). Nason et al. (2000) introduce the *Haar moving average processes*, $\{X_t^{(r)}\}$, of order $2^r - 1$. The simplest Haar MA process with $r = 1$ is

$$X_t^{(1)} = 2^{-1/2}(\epsilon_t - \epsilon_{t-1}), \tag{5.12}$$

where $\{\epsilon_t\}$ is a purely random process with mean zero and variance of σ^2. The process $X_t^{(1)}$ is a LSW process where $w_{j,k;T}$ and $W_j(z)$ are equal to one for $j = -1$ (all k,z), and zero for all other j, $\xi_{-1k} = \epsilon_k$, and $\psi_{j,k}(t)$ are just the Haar discrete non-decimated wavelets. The second-order Haar MA process is

$$X_t^{(2)} = (\epsilon_t + \epsilon_{t-1} - \epsilon_{t-2} - \epsilon_{t-3})/2 \tag{5.13}$$

and is a LSW process according to representation (5.10) with $\psi_{j,k}(t)$ as before, $\xi_{-2k} = \epsilon_k$, and $w_{j,k;T} = 1$ for $j = -2$, and zero otherwise, and so on for higher orders of r. We shall return to this example later. Note how the process coefficients in (5.12) and (5.13) are just those of the finest- and next finest-scale discrete Haar wavelet coefficients. □

In the general time series model in (5.4) the amplitude $A(\omega)$ controls the 'volume' of the sinusoidal oscillation at frequency ω. The usual 'summary statistic' is the *spectrum*, which is $f(\omega) = |A(\omega)|^2$. A similar quantity is defined for the LSW case: the *evolutionary wavelet spectrum* (EWS), $S_j(z)$, is defined as

$$S_j(z) = |W_j(z)|^2, \tag{5.14}$$

for $j = -1, -2, \ldots, -J(T)$, $z \in (0,1)$. The EWS determines how power (variance, to be made more precise later) is distributed across scale j and location $z \in (0,1)$. The quantity z is known as *rescaled time* and a useful working definition is $z = k/T$. The concept of rescaled time was introduced by Dahlhaus (1997), and its use in LSW processes is rather subtle and deserves careful consideration. In particular, the process definition in (5.10) does not define just one stochastic process but a triangular array of process*es*, one for each T. This observation is important when one wishes to consider forecasting LSW processes, see Section 5.4.

5.3.3 LSW simulations

Usually, when one has a potentially useful time series model, it is valuable to be able to simulate from it. The following shows how to use WaveThresh to simulate an LSW process. First, we need to decide two things, (i) the length of the proposed series, T, and (ii) the $S_j(z)$ the EWS underlying the series. For the latter we must specify $S_j(z)$ for $j = -1, \ldots, -J$, where $J = \log_2(T)$ for all $z \in (0,1)$. In WaveThresh we specify the spectrum in actual time, i.e. for values $z = k/T$, where $k = 1, \ldots, T$.

For example, suppose we wanted $T = 1024$ and the spectrum $S_j(z)$ to be specified as

$$S_j(z) = \begin{cases} \sin^2(4\pi z) & \text{for } j = -6, \ z = (0,1), \\ 1 & \text{for } j = -1, \ z \in (800/1024, 900/1024), \\ 0 & \text{else.} \end{cases} \tag{5.15}$$

It is important to remember that spectra must be non-negative.

In WaveThresh EWS are stored as non-decimated wavelet transform objects (class wd and type of station). To construct the above spectrum, one first creates an empty EWS by using the function (cns)

```
> myspec <- cns(1024)
```

Then we have to fill in the spectral contents. First, scale level $j = -6$ is six scale levels away from the finest scale, so in WaveThresh level notation we have to fill level $\log_2(1024) - 6 = 4$ with the squared sinusoid:

```
> myspec <- putD(myspec,level=4,
+     sin(seq(from=0, to=4*pi, length=1024))^2)
```

Then we have to install a 'burst' at the finest scale (level nine in WaveThresh notation) by

```
> burstat800 <- c(rep(0,800), rep(1,100), rep(0,124))
> myspec <-putD(myspec, level=9, v=burstat800)
> plot(myspec, main="", sub="", ylabchars=-(1:10),
+     ylab="Scale")
```

The last plotting command produces the spectral plot shown in Figure 5.5.

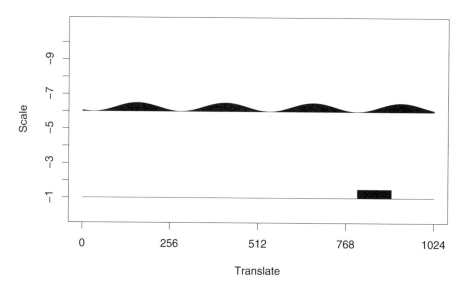

Fig. 5.5. Square-sine at scale -6 and 'burst at 800' at scale -1 EWS, $S_j(z)$ as defined in the text. Produced by f.tsa3().

The object myspec contains the specification for our desired EWS. To simulate a process, we now just use the LSWsim function as follows:

```
> myproc <- LSWsim(myspec)
> ts.plot(myproc, ylab="myproc, X_t")
```

and the plotting function produces a picture of a realization from the spectrum in Figure 5.6. The general sinusoidal nature of the spectrum at a fairly coarse scale (scale -6) can be seen in most of the plot and the very high frequency content due to the burst between 800 and 900 can also be clearly seen. We encourage the reader to try to construct a multiburst spectrum and the associated realization associated with Figure 2 in Nason et al. (2000).

The LSWsim() function takes a direct approach to simulation. It takes the (positive) square root of the supplied EWS spectrum and constructs the

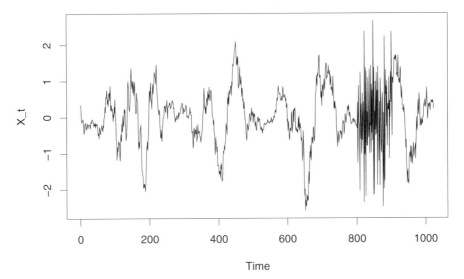

Fig. 5.6. Realization from the square sine and 'burst at 800' spectrum. Produced by `f.tsa4()`.

coefficients $w_{j,k}\xi_{j,k}$, where $w_{j,k} = S_j(k/T)^{1/2}$ and the $\xi_{j,k}$ used are Gaussian with zero mean and unit variance.

5.3.4 Local autocovariances and autocorrelation wavelets

The LSW model says that $X_{t,T}$ is a linear combination of oscillatory waveforms; not surprisingly the autocovariance function of $X_{t,T}$ is one too. However, since $X_{t,T}$ is locally stationary, then clearly the autocovariance function cannot be the ordinary one of stationary time series theory. Since the statistical properties of a locally stationary series vary with time, this must mean that the autocovariance function (if it exists) c must also depend on time, i.e., $c(z, \tau)$.

Thinking back to representation (5.10) one can see that if we want the autocovariance function of $X_{t,T}$, then we will also need to know about the autocovariance/correlation function of the representing wavelets due to the linearity of covariance. Hence, Nason et al. (2000) define the *autocorrelation wavelets*, $\Psi_j(\tau)$, of the discrete wavelets by

$$\Psi_j(\tau) = \sum_k \psi_{j,k}(0)\psi_{j,k}(\tau), \qquad (5.16)$$

for all $j < 0$ and $\tau \in \mathbb{Z}$.

As an example for the Haar wavelets Nason et al. (2000) demonstrate that the *continuous-time* Haar autocorrelation wavelet is given by

$$\Psi_H(u) = \int_{-\infty}^{\infty} \psi_H(x)\psi_H(x-u)\,dx = \begin{cases} 1 - 3|u| & \text{for } |u| \in [0, 1/2], \\ |u| - 1 & \text{for } |u| \in (1/2, 1], \end{cases} \quad (5.17)$$

where $\psi_H(x)$ is the continuous-time Haar mother wavelet from (2.38). One can obtain discrete autocorrelation wavelets by the formula $\Psi_j(\tau) = \Psi(2^j|\tau|)$, where the former Ψ_j is the discrete and the latter Ψ is the autocorrelation function of the associated continuous-time wavelet. The `WaveThresh` package contains a function, `PsiJ`, which computes discrete autocorrelation wavelets. Figure 5.7 shows some examples of autocorrelation wavelets computed with `PsiJ`.

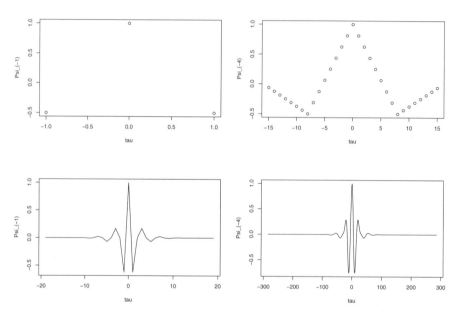

Fig. 5.7. Some discrete autocorrelation wavelets. *Top row*: Haar autocorrelation wavelet: (*left*) $\Psi_{-1}(\tau)$, (*right*) $\Psi_{-4}(\tau)$. *Bottom row*: Daubechies' extremal-phase wavelet $D10$: (*left*) $\Psi_{-1}(\tau)$, (*right*) $\Psi_{-4}(\tau)$. Note that the discrete autocorrelation wavelets are only defined at the integers. We have used a continuous looking line plot instead of little circles for the $D10$ wavelet to make the structure of the function clearer. Produced by `f.tsa5()`.

Nason et al. (2000) show that the autocovariance of $X_{t,T}$ defined by

$$c_T(z, \tau) = \text{cov}(X_{\lfloor zT \rfloor, T}, X_{\lfloor zT \rfloor + \tau, T}) \quad (5.18)$$

converges to $c(z, \tau)$ as $T \to \infty$, where

$$c(z, \tau) = \sum_{j=-\infty}^{-1} S_j(z)\Psi_j(\tau), \quad (5.19)$$

and $\lfloor x \rfloor$ is the largest integer less than or equal to x. Formula (5.19) is a beautiful link between the (limit of the time-varying) autocovariance of $X_{t,T}$ to its (evolutionary wavelet) spectrum by a kind of 'wavelet' transform (the autocorrelation wavelet). Indeed, (5.19) is the analogue of the usual formula, which says that the autocovariance of a stationary process is the Fourier transform of its spectrum.

5.3.5 LSW estimation

For practical time series analysis it is important to possess an effective means of estimating important quantities. Nason et al. (2000) define the *raw wavelet periodogram* as a device to estimate the EWS. The problem is that, given a time series $x_1, \ldots, x_T$, we wish to estimate the EWS of the underlying stochastic process. The fact that we can do this with a single stretch of time series data is guaranteed by the rescaled time device due to Dahlhaus (1997).

Informally speaking, (5.10) shows that X_t is the 'inverse wavelet transform' of the coefficients $w_{j,k;T}\xi_{j,k}$. Thus, our first step in the estimation of $S_j(z)$ is to do the reverse: perform the forward wavelet transform on $\{x_t\}$. Since (5.10) is a non-decimated transform, we need to take the non-decimated wavelet transform of $\{x_t\}$ and this will give us $w_{j,k;T}$ (or $W_j(z)$). Since the EWS is the square of W_j, we can estimate the EWS by the square of the non-decimated wavelet coefficients of $\{x_t\}$.

Thus, Nason et al. (2000) define the empirical non-decimated wavelet coefficients of x_t by

$$d_{j,k;T} = \sum_{t=1}^{T} x_t \psi_{j,k}(t),\tag{5.20}$$

and the (raw) wavelet periodogram is given by

$$I_{k,T}^{j} = |d_{j,k;T}|^2.\tag{5.21}$$

They then show (Proposition 4) that if $\mathbf{I}(z) := \left\{ I_{\lfloor zT \rfloor, T}^{\ell} \right\}_{j=-1,\ldots,-J}$ is the vector of raw periodograms (at z), then

$$\mathbb{E}\left\{\mathbf{I}(z)\right\} = A\mathbf{S}(z) + O(T^{-1}),\tag{5.22}$$

for all $z \in (0,1)$, where $\mathbf{S}(z) = \{S_j(z)\}_{j=-1,\ldots,-J}$, and where matrix A is defined by

$$A_{j\ell} = \langle \Psi_j, \Psi_\ell \rangle = \sum_{\tau} \Psi_j(\tau)\Psi_\ell(\tau),\tag{5.23}$$

which is the inner product matrix of the autocorrelation wavelets. Eckley and Nason (2005) use the multiscale nature of (discrete) wavelets to show how fast algorithms can be used to construct A.

Formula (5.22) shows that the raw wavelet periodogram is, on the average, a blurred or biased version of the spectrum, S (which we wish to estimate). So we need to know A so that we can deblur our estimate of S.

The WaveThresh package contains a function called ipndacw which computes A for various different wavelet families. For example, for Haar wavelets for $J = -6$ the matrix is

```
> ipndacw(-6, 1, family="DaubExPhase")
           -1        -2        -3        -4         -5         -6
-1  1.500000  0.750000  0.375000  0.187500   0.093750   0.046875
-2  0.750000  1.750000  1.125000  0.562500   0.281250   0.140625
-3  0.375000  1.125000  2.875000  2.062500   1.031250   0.515625
-4  0.187500  0.562500  2.062500  5.437500   4.031250   2.015625
-5  0.093750  0.281250  1.031250  4.031250  10.718750   8.015625
-6  0.046875  0.140625  0.515625  2.015625   8.015625  21.359375
```

Nason et al. (2000) show for Haar that this matrix is

$$A_{jj} = \frac{2^{2j} + 5}{3 \times 2^j}, \tag{5.24}$$

$$A_{j\ell} = \frac{2^{2j-1} + 1}{2^\ell}, \tag{5.25}$$

the latter for $\ell > j > 0$ only—the A matrix is symmetric but Formula (5.25) is not.

For a smoother wavelet, say Daubechies' extremal-phase wavelet with ten vanishing moments, the matrix, again for $J = -6$, is

```
> round(ipndacw(-6, 10, family="DaubExPhase"),3)
       -1     -2     -3      -4      -5      -6
-1  1.839  0.322  0.000   0.000   0.000   0.000
-2  0.322  3.035  0.643   0.001   0.000   0.000
-3  0.000  0.643  6.070   1.285   0.002   0.000
-4  0.000  0.001  1.285  12.141   2.570   0.003
-5  0.000  0.000  0.002   2.570  24.282   5.140
-6  0.000  0.000  0.000   0.003   5.140  48.563
```

The matrix becomes 'more' diagonal as the number of vanishing moments of the underlying wavelet increases. However, sadly, apart from two special cases, no closed-form formula is known for A for Daubechies compactly wavelets. One special case is for the Haar wavelet above. The other special case is for the Shannon wavelet, see Section 2.6.1, where A is diagonal with $A_{jj} = 2^{-j}$ for $j < 0$.

Formula (5.22) shows that $\mathbf{I}(z)$ is a biased estimator of $\mathbf{S}(z)$. However, after an obvious bias correction of A^{-1} is applied, the *corrected* wavelet periodogram is

$$\mathbf{L}(z) = A^{-1}\mathbf{I}(z). \tag{5.26}$$

Then Nason et al. (2000) show that

$$\mathbb{E}\left\{\mathbf{L}(z)\right\} = \mathbf{S}(z) + O(T^{-1}). \tag{5.27}$$

The bias correction is very important, as can be seen in Figures 5.8 and 5.9. These two figures produce similar cartoons to those in Section 5.2.1, except

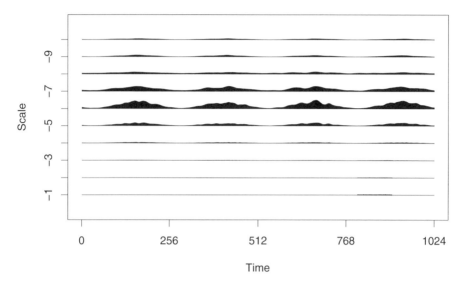

Fig. 5.8. Mean of 100 *uncorrected* periodogram estimations computed on realizations from spectrum shown in Figure 5.5. Produced by `f.tsa6()`.

now for locally stationary processes. The process that produces these figures begins with the spectrum shown in Figure 5.5 and simulates a realization from it. Then we compute the raw wavelet periodogram in (5.21) (for Figure 5.8) and the corrected wavelet periodogram in (5.26) (for Figure 5.9) to estimate the spectrum. We repeat this process for 100 realizations and then average the respective periodograms. The results with the corrected wavelet periodogram in Figure 5.9 are much better than for Figure 5.8. The effect of the application of the bias correction, A^{-1}, is two-fold: first the non-diagonal matrix multiplication deblurs the estimate at one scale from contributions from other scales, and second, a scale factor of about 2^j is applied to scale j so that the fine scales get scaled up relative to the coarse scales.

The other feature of Figures 5.8 and 5.9 is that the raw wavelet periodogram is all non-negative (since it is a squared quantity, see (5.21)). However, the corrected wavelet periodogram is not necessarily non-negative and, indeed, some small negativity is apparent in Figure 5.9. Since the spectrum, $S_j(z)$, itself is non-negative, it would be useful to be able to construct good estimates that are also non-negative, see Fryzlewicz (2003, Section 4.4.3) for

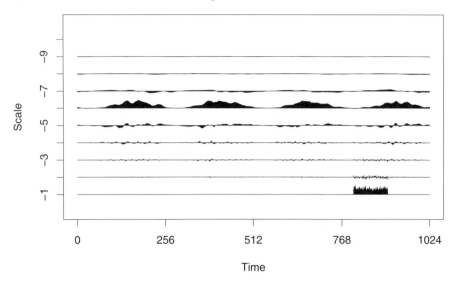

Fig. 5.9. Mean of 100 *corrected* periodogram estimations computed on realizations from spectrum shown in Figure 5.5. Produced by `f.tsa7()`.

further details of this through formulation as a linear complementarity problem.

As well as the expectation of the raw wavelet periodogram given in (5.22), Nason et al. (2000) derive the following formula for the variance:

$$\text{var}(I^j_{\lfloor zT \rfloor, T}) = 2 \left\{ \sum_\ell A_{j\ell} S_\ell(z) \right\}^2 + O(2^{-j}/T), \qquad (5.28)$$

for all $z \in (0, 1)$. Note that this means that the variance of the wavelet periodogram does not vanish as the sample size, $T \to \infty$. Hence, like the situation in the stationary case, the wavelet periodogram is not a consistent estimator of the EWS. To solve this problem Nason et al. (2000) adopt the well-tried remedy (from the stationary situation) of smoothing the wavelet periodogram as a function of $\lfloor zT \rfloor$ (or k) for each scale j. Then they recommend applying the correction to the smoothed wavelet periodogram. It would be possible to correct first and then smooth, but then this is more difficult to analyze from a theoretical point of view. Nason et al. (2000) recommend the use of wavelet shrinkage to smooth the wavelet periodogram.

5.3.6 More on smoothing the wavelet periodogram

More recent work by Fryzlewicz (2003), Fryzlewicz and Nason (2003), and Fryzlewicz and Nason (2006) advocates an approach using wavelet-Fisz transforms, see Chapter 6.

5.3.7 LSW analysis of Baby ECG data

Let us now return to the `BabyECG` data from the beginning of the chapter. We use the following commands in `WaveThresh` to produce the corrected wavelet periodogram estimate of $S_j(z)$ shown in Figure 5.11.

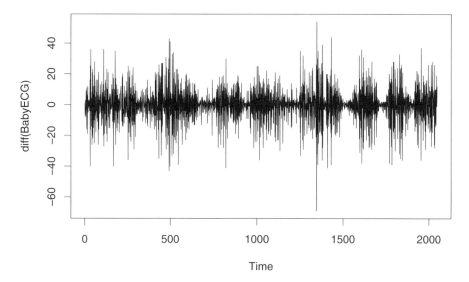

Fig. 5.10. Differenced `BabyECG` data. Produced by `f.tsa8a()`.

Note that we operate on the differenced data depicted in Figure 5.10, so that the series has approximately zero mean as required by the LSW specification in (5.10).

```
# Start time (in hours)
> sttm <- 21+(17+59/60)/60

# Labels for x axis
> tchar <- c("22", "23", "00", "01", "02", "03", "04", "05",
+       "06")
# Numerical values for x axis, convert them to x values
> tm2 <- c(22,23, 24, 25, 26, 27, 28, 29, 30)
> tm2 <- tm2 - sttm
# Convert to seconds
> tm2s <- tm2*60*60
# Convert to sampling units
> tm2u <- tm2s/16

# Plot the differenced data
#
```

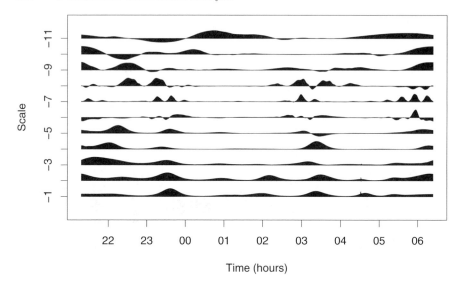

Fig. 5.11. Corrected smoothed wavelet periodogram of the differenced `BabyECG` data. Both LSW discrete wavelets and smoothing wavelets are Daubechies' least-asymmetric wavelets with ten vanishing moments. Each periodogram was level smoothed by log transform, followed by translation-invariant global universal thresholding with MAD variance estimation on all smoothed levels (4:10), followed by inverse exp transform. All levels are individually scaled to show maximum detail. Produced by `f.tsa8()`.

```
> ts.plot(diff(BabyECG))

# Now prepare differenced data for analysis (add on an
# observation at the front so as to make the differenced
# series an appropriate length).

> dBabyECG <- diff(c(BabyECG[2], BabyECG))

# Compute corrected smoothed wavelet periodgram with  options
> spec <- ewspec(dBabyECG,
+        smooth.levels=4:(nlevels(BabyECG)-1),
+        smooth.policy="universal", smooth.transform=log,
+        smooth.inverse=exp)$S

# Plot the estimate
> plot(spec, main="", sub="", ylabchars=-(1:11),
+        scaling="by.level", ylab="Scale",
+        xlab="Time (hours)",
+        xlabvals=tm2u, xlabchars=tchar)
```

All the commands up to the `ewspec()` call are involved in setting up the horizontal time axis correctly. The corrected smoothed wavelet periodogram is calculated by the `ewspec()` function. The `ewspec()` function is a complicated function with many options that can be specified. All options for `ewspec()` that begin with `smooth.` have to do with the smoothing of the raw periodogram. In the example above, each scale level from the raw wavelet periodogram is subjected to a log transform (`smooth.transform=log`), then ordinary universal wavelet thresholding (`smooth.policy="universal"`, described in Chapter 3) is used, and then the inverse transform (exp) is applied (`smooth.inverse=exp`). This transform-smooth-inverse method does introduce a bias. Also, for each scale of the wavelet periodogram only levels four and finer are thresholded (`smooth.levels=4:(nlevels(BabyECG)-1)`). Scales four and finer using `WaveThresh` indexing correspond to scale levels -7 and finer, using Nason et al. (2000) indexing.

The idea behind using a log transform is that (for Gaussian data) the distribution of the raw wavelet periodogram is approximately χ^2 and the use of log stabilizes variance and draws the distribution towards normality, and thereby permitting universal thresholding, which is designed to work in this situation. Alternative methods for smoothing using variance stabilization techniques can be found in Chapter 6. The extra `plot()` arguments have mostly to do with changing the values of the horizontal and vertical axes to display time of observation and LSW.

As mentioned in Section 5.1, the `BabyECG` time series comes with another, the `BabySS` sleep state series. As a practical means for assessing the potential utility of LSW processes and the associated estimation methodology it is useful to plot both the spectral estimate (or one scale of it) and the sleep state on the same graph. Figure 5.12 shows our estimate of $S_{-1}(z)$: the contribution to variance over time at the finest scale. It is clear that the finest-scale power correlates fairly closely with the sleep state (and indeed the other fine scales, e.g., -2, also correlate well with the sleep state). In this plot, high power corresponds to the infant being awake and low power to being asleep.

Note that the medium and coarser scales do not correlate as well with the sleep state, and so a scale analysis has taught us something. A simple local variance estimator also correlates well with the sleep state, and an appropriately windowed local Fourier-based estimator would also probably do so as well. Nason et al. (2001) perform a more general analysis using nondecimated wavelet packets to find good predictors of sleep state from heart rate, and it turns out that ordinary father wavelets, averaging over different dyadic periods of time, do the best. This particular example, using both time series, is most useful as the sleep state variable verifies that the multiscale LSW EWS spectral estimate is a useful quantity, although not uniquely so.

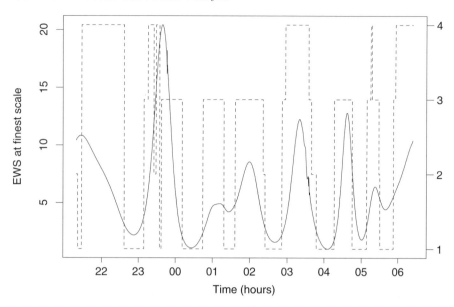

Fig. 5.12. *Solid line*: estimate of $S_{-1}(z)$ (corrected smoothed wavelet periodogram at scale -1) for the `BabyECG` data—values on the left-hand axis. *Dashed line*: sleep state as determined by expert analysis—values on the *right-hand axis*: (1=quiet sleep, 2=state between 1 and 3; 3=active sleep; 4=awake). Reproduced from Nason et al. (2000) with permission.

5.3.8 LSW analysis of RSAM data

Figure 5.13 shows the corrected smoothed wavelet periodogram for the RSAM data introduced in Section 5.1. Since the sampling frequency of the RSAM time series is one observation every 5 minutes, the Nyquist frequency (the highest observable frequency in the time series) is one complete oscillation every 10 minutes or a period of 10 minutes (see, e.g., Chatfield (2003)). The finest-scale EWS approximately captures variation in the series in the highest half of the frequencies: this is periods 20 minutes to 10 minutes; the mid-period of this range is 15 minutes. The next finest-scale EWS (-2) approximately contains periods in the range of $[20, 40]$ minutes and so the mid-period is 30, and so on. The vertical labels in Figure 5.13 show the mid-periods for each scale band. Note that the top scale band shown corresponds to periods of 85 days. However, the length of the time series only corresponds to 60 days, so how can we observe oscillations of 85 days? We cannot, and in these spectral plots we should not, take too much notice of the coarsest-scale estimate, especially near the ends (with `WaveThresh` the transforms used here are periodic and so the ends 'interfere'). The curved-dashed lines are drawn to indicate that only the spectral estimate below both curved lines is to be interpreted. These lines are similar to the 'cone of influence' described by Torrence and Compo (1998), who describe another multiscale method of

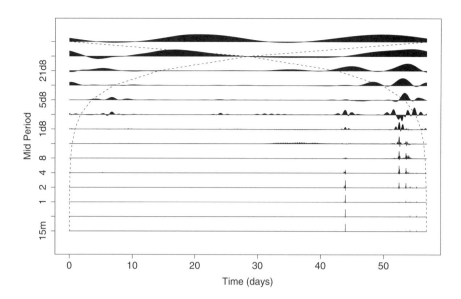

Fig. 5.13. Corrected smoothed wavelet periodogram of the RSAM data shown in Figure 5.2. Both LSW discrete wavelets and smoothing wavelets are Daubechies' least-asymmetric wavelets with ten vanishing moments. Each periodogram level smoothed by translation-invariant universal thresholding with (`LSuniversal`) threshold of $\hat{\sigma}\sqrt{\log(T)}$ on levels 3:13. All levels are individually scaled to show maximum detail. Reproduced from Nason (2006) with permission from the International Association of Volcanology and Chemistry of the Earth's Interior.

spectral analysis using wavelet methods. Unlike the `BabyECG` data, the author does not have access to a 'covariate' time series for RSAM, but with data of this sort there often is covariate information (for example, an earthquake is either happening or not!) However, some interpretations, largely supplied by Willy Aspinall, can be made and are described in Nason (2006) as follows. A large peak exists in the series at about 43 days (Sept 25th, 2001), and this can be clearly determined in the spectral estimate as a high frequency feature (indeed, the dominant feature at these frequencies). Within the 8-hours to 2-day-16-hour bands there appears to be oscillation at 52 days that might reflect a phenomenon known as 'banded tremor'. However, as pointed out by Nason (2006), this is a preliminary analysis and much further work on interpretation and linking to real events still needs to be carried out.

5.4 Forecasting with Locally Stationary Wavelet Models

The previous sections have introduced a time series model with a time-varying spectrum, $S_j(z)$, along with an associated time-localized covariance, $c(z, \tau)$. If one has an observed time series $X_{0,T}, X_{1,T}, \ldots, X_{t-1,T}$, our methods above allow us to model and estimate $S_j(z)$ and $c(z, \tau)$.

A key problem in time series analysis is that if one has such an observed series $X_{0,T}, \ldots, X_{t-1,T}$ then is it possible to forecast, or predict, future values of X, and, if so, how?

In a remarkable paper, Fryzlewicz et al. (2003) show how forecasting can be carried out using a time series modelled by a LSW process. The rest of this section describes their ideas.

Suppose one observes the sequence $X_{0,T}, \ldots, X_{t-1}$. Fryzlewicz et al. (2003) define the *h-step-ahead predictor* of $X_{t-1+h,T}$ by

$$\hat{X}_{t-1+h,T} = \sum_{s=0}^{t-1} b_{t-1-s;T}^{(h)} X_{s,T}, \qquad (5.29)$$

where the coefficients $b_{t-1-s;T}^{(h)}$ are chosen to minimize the 'usual' mean square prediction error defined by

$$\text{MSPE}(\hat{X}_{t-1+h,T}, X_{t-1+h,T}) = \mathbb{E}\left\{\left(\hat{X}_{t-1+h,T} - X_{t-1+h,T}\right)^2\right\}. \qquad (5.30)$$

Although, of course, here it is not quite the usual coefficients $b_{t-1-s;T}^{(h)}$ because in the stationary theory the coefficients do not depend on T; more on this later.

In their set-up they assume that $T = t + h$, i.e., T pieces of observed data are potentially available and that only the first t get observed. So if observations at times $0, \ldots, t-1$ are observed, then a one-step prediction would predict X_t, and the prediction farthest out predicts $X_{T-1,T}$. Using this notation the last observation is $X_{t-1,T} = X_{T-h-1,T}$. Hence, with all this their methodology can build an estimate for the EWS, $S_j(z)$ only on the rescaled time interval

$$\left[0, 1 - \frac{h+1}{T}\right], \qquad (5.31)$$

as that is where the observed data map to. The *predicted* values of $S_j(z)$ will be those on the rescaled time interval

$$\left(1 - \frac{h+1}{T}, 1\right). \qquad (5.32)$$

As $T \to \infty$, the estimation domain expands to $(0, 1)$ and the prediction domain shrinks to the empty set. So as more and more data get collected we eventually get to know the whole spectrum, and nothing need be predicted. Fryzlewicz et al. (2003) remark that the predictor (5.29) can be written in terms of a non-decimated wavelet expansion with a different set of prediction coefficients. However, this latter set of prediction coefficients is not unique due to the overdetermined nature of the non-decimated wavelets (just as the $w_{j,k;T}$ coefficients in the LSW representation in (5.10)).

For one-step-ahead prediction Fryzlewicz et al. (2003) show that the MSPE may be approximated by $b_t^T B_{t;T} b_t$, where b_t is the vector $\left(b_{t-1;T}^{(1)}, \ldots, b_{0;T}^{(1)}, -1 \right)$ and $B_{t;T}$ is a $(t+1) \times (t+1)$ matrix whose (m,n)th element is given by

$$\sum_{j=-J}^{-1} S_j \left(\frac{n+m}{2T} \right) \Psi_j(n-m), \tag{5.33}$$

which can be estimated using an estimate for $S_j(z)$. They then show that the minimizing $\{b_{s;T}^{(1)}\}$ can be found by solving the linear system given by

$$\sum_{m=0}^{t-1} b_{t-1-m;T}^{(1)} \left\{ \sum_{j=-J}^{-1} S_j \left(\frac{n+m}{2T} \right) \Psi_j(m-n) \right\} = \sum_{j=-J}^{-1} S_j \left(\frac{n+t}{2T} \right) \Psi_j(t-n), \tag{5.34}$$

for each $n = 0, \ldots, t-1$. Using Formula (5.19), which links the spectrum and localized autocovariance, the prediction coefficient Equations (5.34) can be written in terms of localized autocovariances as:

$$\sum_{m=0}^{t-1} b_{t-1-m;T}^{(1)} c \left(\frac{n+m}{2T}, m-n \right) = c \left(\frac{n+t}{2T}, t-n \right). \tag{5.35}$$

It is entirely satisfactory to note that in the special case where X_t is also a stationary process, the localized autocovariance loses its dependence on (rescaled) time z and becomes equal to the ordinary autocovariance function $c(z, \tau) = c(\tau)$ and the prediction equations in (5.35) become

$$\sum_{m=0}^{t-1} b_{t-1-m}^{(1)} c(m-n) = c(t-n), \tag{5.36}$$

again for $n = 0, \ldots, t-1$, and these latter equations are the standard Yule-Walker equations used to forecast stationary processes, see, for example, Brockwell and Davis (1991).

Fryzlewicz et al. (2003) also develop a nice analogue of the Kolmogorov formula for the theoretical one-step prediction error (a modification of a result from Dahlhaus (1996) for locally stationary Fourier processes) and consider h-step-ahead prediction.

An important aspect of Fryzlewicz et al. (2003) is in the development of methodology, algorithms, and code to enable prediction from given data. They propose to solve the (localized) Yule–Walker equations in (5.35) using estimates of the localized autocovariance. Their estimates of the localized autocovariance are obtained by using Formula (5.19) and estimating the spectrum $S_j(z)$ by the corrected wavelet periodogram **L** given in (5.26). As with the corrected wavelet periodogram, their localized autocovariance estimates need to be smoothed and they use standard kernel smoothing to accomplish this. They further define a 'clipped' predictor, $\hat{X}_{t,T}^{(p)}$, which is the predictor given in (5.29), except the summation starts at $s = t - p$. In other words, the clipped predictor uses the last p observations to predict. The reason for this is that, overall, for every extra T, the estimation error increases with each extra component of the b_t prediction coefficient vector. On the other hand, the prediction error decreases as more observations are recorded, and the clipped predictor achieves a balance between these two competing criteria. Fryzlewicz et al. (2003) then use an iterative algorithm to choose good values of p and the kernel smoothing bandwidth.

5.4.1 Using LSW forecasting software

Some sample code that implements the Fryzlewicz et al. (2003) forecasting technique is available from Piotr Fryzlewicz's website:

http://www.stats.bris.ac.uk/~mapzf .

The two main functions to use are called adjust, which calculates good values of p and the kernel smoothing bandwidth, and pred, which uses those 'good' values to predict future values.

Forecasting simulated series. For example, let us return to the simulated time series generated in Section 5.3.3 on p. 179 and depicted in Figure 5.6. The series is of length 1024. Let us produce two forecasts. In each case, we shall use part of the simulated series, up to time $t - 1$, to build a forecast, and then forecast the tth value. First, let $t = 1024$. Then, to build the forecast on the observations up to $t - 1$, we execute the following code:

```
> myproc1 <- myproc[1:1023]
> myproc1.par <- adjust(myproc1) # Calculate good p and g
> myproc1.pred <- pred(myproc1,
+      p = myproc1.par$p[21], g = myproc1.par$g[21])
```

The predicted value is stored in the $mean component of myproc1.pred and its standard error in the $sd component. For this realization, the predicted value turns out to be exactly zero, and its standard error is 0.624. The true value (which we know in the case) is 0.221. Thus, the prediction is not too far away from the truth, and well within one standard deviation. We have performed this prediction in an area where the spectrum consists of relatively low frequencies. We can repeat the exercise for $t - 1 = 850$ and predict the 851st value. In this area the spectrum has both the low-frequency oscillations at scale -6, but also the very high-frequency burst at scale -1. If we repeat the exercise, the predicted value is 1.37 with standard error of 0.949. The true value is 0.146. So this time the prediction is within two standard errors, but not one.

Forecasting Baby ECG series. Figure 5.14 shows the LSW forecasting methodology applied to the `BabyECG` data from earlier. The commands we used to produce this figure were as follows. First, we generated a new time series, `pBaby`, that contained the series up to time point $t = 2038$. We ran the prediction on this series and attempted to predict values of the series beyond that. The variable `nsteps` is the number of steps that we want to predict beyond 2038, in this case ten. The variable `s` is a parameter for `adjust()` that tells it how far back to start the method that figures out good p and bandwidth parameters.

```
> nsteps <- 10
> s <- 40
> nBaby <- length(BabyECG)
> pBaby <- BabyECG[1:(nBaby - nsteps)]
```

We also 'mean-correct' the baby series so that the mean of the series that we forecast from is zero, as must be the case for LSW processes.

```
> meanpBaby <- mean(pBaby)
> pBaby <- pBaby - meanpBaby
```

Now we create objects that will store our predicted values (and, incidentally, their standard errors, although Figure 5.14 does not use them).

```
> ppred <- rep(0, nsteps)
> pse <- rep(0,nsteps)
```

And now follows the key part of the algorithm, which, for each prediction ahead, runs the `adjust()` function to choose good p and bandwidth parameters and then calls `pred()` to do the prediction.

```
> for(h in 1:nsteps)        {
+            pBaby.par <- adjust(pBaby, h=h, s=s)
+            pBaby.pred <- pred(pBaby, h=h,
+                    p = pBaby.par$p[s+1], g = pBaby.par$g[s+1])
+            ppred[h] <- pBaby.pred$mean + meanpBaby
+            pse[h] <- pBaby.pred$std.err
+            }
```

The next few commands figure out a decent vertical range to use if we intend to plot both the true function and the forecasts. The true time series and the forecast values are plotted in Figure 5.14 using the R `points()` function

```
# BebeVals are just the ones we want to plot
> BebeVals <- (nBaby - 2*nsteps+1):nBaby

> yr <- range(c(BabyECG[BebeVals], ppred))
```

```
> plot(BebeVals, BabyECG[BebeVals],
+  type="l", ylim=yr, xlab="Observation", ylab="BabyECG")
+ points(BebeVals, ppred)
```

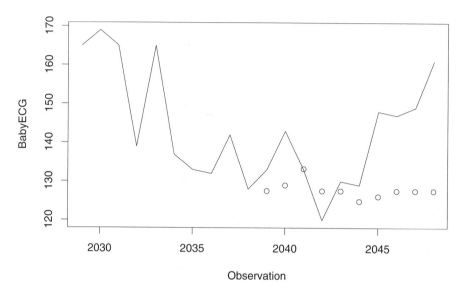

Fig. 5.14. LSW forecasting on `BabyECG` time series depicted in Figure 5.1. The predictions are based on observations up to time 2038. The *small circles* show the predictions from times 2039 until 2048. The *solid line* shows the actual series (and hence the true values from 2039 to 2048). Produced by `f.tsa9()`.

5.4.2 Basis for forecasting with LSW non-stationarity

At this point we have not paid too much attention to the stochastic model with respect to the definition of the locally stationary process given around (5.10). However, one should note that the definition does not define a *single* stochastic process but a triangular array consisting of a set of different stochastic processes. To aid understanding we could depict the processes as follows:

$$X_{0,1}$$
$$X_{0,2}, X_{1,2}$$
$$X_{0,3}, X_{1,3}, X_{2,3}$$
$$X_{0,4}, X_{1,4}, X_{2,4}, X_{3,4}$$
$$\vdots$$
$$X_{0,T}, \ldots, X_{t-1,T}$$
$$\vdots$$

However, each process obeys the representation (5.10), which shares the common underlying spectrum $S_j(z)$. An 'exception' to this is the class of stationary processes where actually each row *is* part of the same process, but this does not contradict the LSW model because here the spectrum no longer depends on (rescaled) time z and, indeed, the $w_{j,k}$ actually also do not then depend on k.

At first sight, it may seem that the LSW model might not be suitable for prediction. In the classical stationary case, as a new datum is observed, e.g., T goes to $T+1$, the existing series values remain unchanged and the new datum is treated as new information. However, the 'triangular array' above for the LSW model seems to suggest that as a new observation arrives a whole new realization is activated and possibly the values previously used for building forecasts change!

Fortunately, the latter interpretation is not correct as it fails to take account of the fact that once a realization has been recorded, any forecast that we make is statistically *conditioned* upon that realized set of data. The underlying spectrum for realizations with T and $T+1$ will be the same, but of course any estimated spectrum will be different. However, because of the 'slow evolution' conditions on the spectrum, it is highly probable that any two estimated spectra for similar values of T will be close.

5.5 Time Series with Wavelet Packets

Wavelet packets have also been employed for the study of time series problems. We have already mentioned the estimation of wavelet variance using wavelet packets by Gabbanini et al. (2004) in Section 5.2.2.

Percival and Walden (2000, Chapter 7) is an excellent reference which explains the discrete wavelet packet transform, the best-basis algorithm (as in 2.11), and the 'maximal-overlap' discrete wavelet packet transform (as in 2.12). Chapter 7 of Percival and Walden (2000) also describes the technique of *matching pursuit* due to Mallat and Zhang (1993), which, given a signal, successively searches a dictionary of waveforms for the best match, computes the residual from this match, and then searches the residual for the next best match, and so on. Matching pursuit is closely related to projection pursuit regression introduced by Friedman and Stuetzle (1981).

Nason and Sapatinas (2002) describe a methodology for building transfer function models between two time series: an explanatory series X_t and a response series Y_t. First, the non-decimated wavelet packet transform of X_t is performed, which generates a large set of candidate packet time series. Each candidate is the projection of X_t onto a set of non-decimated wavelet packet functions. Then each candidate is 'correlated' with the response series Y_t and the best candidates selected to form a candidate group. Then Y_t is regressed in an appropriate way onto the candidate group variables. This regression relationship can then be used in a predictive mode where future values of

X_t, after selecting the appropriate wavelet packet basis elements, and feeding through the regression relationship, can be used to predict future values of Y_t. There are similarities between this technique and matching pursuit, projection pursuit regression (both mentioned above), and techniques such as principal components analysis (PCA) regression (except the PCA is replaced by wavelet packets). This technique can be useful where it is expensive to collect Y_t but cheap to collect X_t as a predictive model can be built where accurate values of Y_t can be predicted merely from values of X_t. Hunt and Nason (2001) use this wavelet packet transfer methodology to generate a predictive model between Met Office wind speed measurements taken at established weather stations, X_t, and those taken at a prospective wind farm site, Y_t. Then historically established station data are fed through the predictive model to generate corresponding Y_t values, which can then be used to predict future wind energy characteristics for the prospective site. Nason et al. (2001) consider another situation where X_t is the heart rate (easy to measure) and Y_t is the sleep state of infants (expensive). The transfer function models then developed are used to classify infant sleep state for future time periods solely from 'cheap' heart-rate data.

5.6 Related Topics and Discussion

For a stationary time series the Fourier basis representation given in (5.4) is canonical. Section 5.3 has shown that for locally stationary time series the locally stationary wavelet process model (5.10) is *but one* potential model. Other fine possibilities include the time-varying Fourier from Dahlhaus (1997) and the SLEX models of Ombao et al. (2002), for example. Many possibilities can be obtained by slotting in other oscillatory basis functions, and a general way to proceed can be found in Priestley (1983, Section 11.2).

However, a wavelet model might not be the best for a particular time series nor might a Fourier version, and something else might be more suitable. Basis choice, and model choice, for nonstationary time series analysis is very much an open problem. However, these models should be seen as examples arising from a rapidly expanding and exciting body of research concerned with locally stationary time series. For example, the time-varying ARCH processes of Dahlhaus and Subba Rao (2006, 2007) are a different, but related, set of models. For many problems, such general locally stationary models may be overspecified and too complicated, and hence restricted models might be more useful. For example, time-modulated processes, see Fryzlewicz et al. (2003), or piecewise constant spectra, see Fryzlewicz et al. (2006).

It is possible to extend the basic LSW model in a number of ways. For example, Van Bellegem and von Sachs (2008) consider spectra with jump discontinuities rather than the smoothness restrictions on $W_j(z)$, as mentioned just after Equation (5.11). Extensions to 2D lattice processes were developed by Eckley (2001), who also investigated their application to the analysis of texture.

Multiscale Variance Stabilization

Variance stabilization is one of the oldest techniques in statistics, see, for example, Barlett (1936). Suppose one has a random variable X, which has a distribution dependent on some parameter θ and that the variance of X is given by $\text{var}(X) = \sigma^2(\theta)$. If X was Gaussian with distribution $X \sim N(\mu, \sigma^2)$ and the parameter of interest is μ, then it is patently clear that the variance σ^2 *does not* depend on μ. For many other random variables, this is not true. For example, if Y was distributed as a Poisson random variable, $Y \sim \text{Pois}(\lambda)$, then the (only) parameter of interest is λ and the variance $\sigma^2(\lambda) = \lambda$. In other words, the variance depends directly on the parameter of interest, λ, which is also the mean. Hence, we often refer to a 'mean-variance' relationship, and for Poisson variables the variance is equal to the mean.

There are many situations where such a mean-variance relationship can cause problems. In many statistical modelling situations such as simple linear regression, nonparametric regressions, and generalized linear models, the assumption of *homoscedastic* error is commonly made. However, often the variance of the data is not constant and sometimes one observes that the variance is a non-decreasing function of the mean. In these circumstances of *heteroscedastic* variance, one solution is to turn to variance stabilization—that is, find a transformation of the data to make the variance constant.

In practical problems, one is often encouraged to try to 'take logs' or 'take square roots' of the data and then to use residual plots to see whether the transformation has been successful at 'stabilizing' the variance. For Poisson random variables, the concept of 'taking square roots' is well known. See, for example, Barlett (1936) or Anscombe (1948). The result in Anscombe (1948) says that if $Y \sim \text{Pois}(\lambda)$, then the transformed value $Z = \sqrt{Y + c}$ has the following asymptotic expansion for its variance as $\lambda \to \infty$:

$$\text{var}(Z) \approx \frac{1}{4} \left\{ 1 + \frac{3 - 8c}{\lambda} + \frac{32c^2 - 52c + 17}{32\lambda^2} \right\}. \tag{6.1}$$

Hence, the second term can be made to vanish by setting $c = \frac{3}{8}$, which results in

$$\text{var}(Z) \approx \frac{1}{4}\left\{1 + \frac{1}{16\lambda^2}\right\}, \tag{6.2}$$

which becomes close to the constant $\frac{1}{4}$ as $\lambda \to \infty$.

6.1 Why the Square Root for Poisson?

Suppose $X \sim \text{Pois}(\lambda)$. Then for variance stabilization we look for a transformation, g, such that

$$\text{var}\{g(X)\} = K. \tag{6.3}$$

In other words, a g such that the variance of the transformed variable $g(X)$ is a constant, K, and, in particular, does not depend on a given parameter of the distribution, which here is λ.

How can we find such a g? Generally, there is no function that can exactly stabilize the variance. Most methods find a function g such that the variance is approximately stabilized and develop an asymptotic regime such that the variance is stabilized asymptotically. Most methods are based on a stochastic Taylor expansion of $g(X)$ and derive from Barlett (1936) and Anscombe (1948) (and earlier). If X is 'concentrated' around its mean λ, then we can derive the following Taylor expansion for $g(X)$ around λ:

$$g(X) = g(\lambda) + (X - \lambda)g'(\lambda) + (X - \lambda)^2 g''(\lambda)/2 + \mathcal{O}(X - \lambda)^3. \tag{6.4}$$

Our aim is to obtain an approximation for $\text{var}\{g(X)\}$. We can obtain this after first finding approximations for $\mathbb{E}\{g(X)\}$ and $\mathbb{E}\{g^2(X)\}$. Ignoring higher-order terms and then taking expectations of (6.4) gives

$$\mathbb{E}\{g(X)\} \approx g(\lambda) + 0 + \text{var}(X)g''(\lambda)/2. \tag{6.5}$$

The linear term in (6.4) disappears since $\mathbb{E}X = \lambda$. Hence, the square of the expectation of $g(X)$ is approximately

$$[\mathbb{E}\{g(X)\}]^2 \approx g(\lambda)^2 + g(\lambda)g''(\lambda)\text{var}(X), \tag{6.6}$$

ignoring the higher-order $\text{var}(X)^2$ term.

Now the square of (6.4) is

$$g^2(X) \approx g^2(\lambda) + 2g(\lambda)g'(\lambda)(X - \lambda) + g(\lambda)g''(\lambda)(X - \lambda)^2 \tag{6.7}$$
$$+ (X - \lambda)^2 g'(\lambda)^2 + g'(\lambda)g''(\lambda)(X - \lambda)^3 + (X - \lambda)^4 g''(\lambda)^2/4.$$

Taking expectations of this, again using $\mathbb{E}(X - \lambda) = 0$, and ignoring moments of three and higher we obtain

$$\mathbb{E}\{g(X)^2\} \approx g^2(\lambda) + \{g(\lambda)g''(\lambda) + g'(\lambda)^2\}\text{var}(X). \tag{6.8}$$

Hence, our approximation for $\text{var}\{g(X)\}$ is derived by subtracting (6.6) from (6.8) giving

$$\text{var}\left\{g(X)\right\} \approx g'(\lambda)^2 \text{var}(X). \tag{6.9}$$

We want the transformed variance to be constant, K, so substituting (6.9) into (6.3) gives

$$g'(\lambda)^2 \text{var}(X) \approx K, \tag{6.10}$$

and hence this means that our choice of g has to satisfy

$$g'(\lambda) \approx K^{1/2}\text{var}(X)^{-1/2}. \tag{6.11}$$

For the Poisson distribution, if $X \sim \text{Pois}(\lambda)$, then we know that $\text{var}(X) = \lambda$ and hence

$$g'(\lambda) \approx K^{1/2}\lambda^{-1/2}, \tag{6.12}$$

and on integrating gives

$$g(\lambda) \approx K_1\sqrt{\lambda}, \tag{6.13}$$

where K_1 is another constant. Hence, the square-root function is appropriate for Poisson. More refined approximation calculations result in, e.g., the $\sqrt{X+3/8}$ transform of Anscombe, and it is easy to see how the same kind of thing can work for other random variables. We denote the Anscombe transform by the letter $\mathcal{A}$.

6.2 The Fisz Transform

For a given random variable the previous section indicates how one might develop a transform so as to approximately stabilize its variance. The transformations in the previous section are all examples of *diagonal* transforms. That is, given a sequence of random variables with the same distributional form (e.g. Poisson) but differing parameters (e.g. intensity λ), one would apply these variance stabilizing transforms one sequence element at a time, i.e., $Y_i = g(X_i)$ for $i = 1, \ldots, n$. However, with a sequence of random variables one might think of using a multivariate function. In other words, the input to the 'stabilization function' might be the whole sequence you wish to transform $(X_1, \ldots, X_n)$ to obtain stabilized output $(Y_1, \ldots, Y_n)$ but it might not only be X_i that contributes to the stabilization of Y_i. Obviously, the previous diagonal transformations are a special case of these, more general, transformations.

As a step on the way to a multivariate variance stabilizer we refer the reader to the Fisz transform, which appears as the following theorem from Fisz (1955) (the version for the Poisson distribution expressed here is taken from Fryzlewicz and Nason (2004)):

Theorem 6.1. *Let $X_i \sim \text{Pois}(\lambda_i)$ for $i = 1, 2$ and X_1, X_2 are independent. Define the function $\zeta : \mathbb{R}^2 \to \mathbb{R}$ by*

$$\zeta(X_1, X_2) = \begin{cases} 0 & \text{if } X_1 = X_2 = 0, \\ (X_1 - X_2)/(X_1 + X_2)^{\frac{1}{2}} & \text{otherwise.} \end{cases} \tag{6.14}$$

If $(\lambda_1, \lambda_2) \to (\infty, \infty)$ and $\lambda_1/\lambda_2 \to 1$, then $\zeta(X_1, X_2) - \zeta(\lambda_1, \lambda_2) \overset{d}{\to} N(0, 1)$.

The application of the theorem is quite revealing and most useful for variable transformation. The theorem says that if you have two Poisson random variables X_1 and X_2 with intensities λ_1 and λ_2, then for large λ_1 and λ_2, where $\lambda_1 \approx \lambda_2$, one has that $\zeta(X_1, X_2) - \zeta(\lambda_1, \lambda_2)$ is approximately a standard normal random variable and hence is approximately variance stabilized (and also Gaussian with zero mean—both useful additional properties).

The reader might notice that the numerator in ζ in (6.14) is essentially the (finest-scale) Haar wavelet coefficient and the denominator is essentially the (finest-scale) Haar father wavelet coefficient. Hence, this means that if one takes the Haar transform of a sequence of Poisson random variables, then one can divide the finest-scale wavelet coefficients by their corresponding father wavelet coefficients, and the resultant variables will, under the right conditions applying to their intensity, be approximately $N(0, 1)$.

Although the above theorem is asymptotic in nature, the asymptotics 'kick in' surprisingly quickly. For example, let us choose $\lambda_1 = \lambda_2 = 2$. In R we can define the Fisz transform using the following code:

```
> fisz <- function(x,y){

+ ans <- (x-y)/sqrt(x+y)
+ans[(x==0) & (y==0)] <- 0

+ans
+ }
```

and then use the following code to simulate two sets of 1000 Poisson random variables with intensities λ_1 and λ_2. Then we plot the Anscombe- and Fisz-transformed versions:

```
> x <- rpois(1000, lambda=2)
> y <- rpois(1000, lambda=2)
> zeta <- fisz(x,y) - fisz(2,2)

> A <- sqrt(x+ 3/8)

> plot(density(A, from=0), main="a.", xlab="A")
> qqnorm(A, main="b.")
> plot(density(zeta, from=-3, to=3), main="c.",
+     xlab="zeta")
> qqnorm(zeta, main="d.")
```

The results of this code are depicted in Figure 6.1. From the density plots the density estimate of the Fisz density appears to look more Gaussian and more symmetric and have less skew than that of the Anscombe transformation. The Q-Q plot for Fisz looks better, too. It is less stepped, and the left-hand tail looks less elongated compared with the values transformed with the Anscombe transformation.

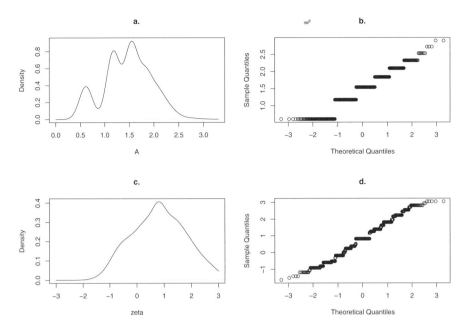

Fig. 6.1. Results of two variance-stabilizing transforms of simulated Poisson data. In both rows the left-hand figure is a kernel density estimate of the transformed variates and the right-hand figure is a Q-Q plot of the same values. The density estimation was carried out using `density()` in R using the `bcv` bandwidth selector. Produced by `f.hf1()`.

For another glimpse of the comparative properties of Anscombe and Fisz let us introduce the following simple simulation that generates Poisson random variables, applies the two stabilizing transforms, and then computes the mean, variance, skewness, and kurtosis on the transformed variables. This simulation is repeated 100 times. First, the function definition:

```
> code.hf1 <- function(){

+ nsims <- 100
+ lam<-2
+ a.mn <- a.vr <- a.sk <- a.ku <- rep(0, nsims)
+ f.mn <- f.vr <- f.sk <- f.ku <- rep(0, nsims)


+ for(j in 1:nsims)         {

+          x <- rpois(1000, lambda=lam)
+          y <- rpois(1000, lambda=lam)

+          zeta <- fisz(x,y)  - fisz(2,2)
```

```
+          f.mn[j] <- mean(zeta)
+          f.vr[j] <- var(zeta)
+          f.sk[j] <- skewness(zeta)
+          f.ku[j] <- kurtosis(zeta)

+          A <- 2*sqrt(x + 3/8) - 2*sqrt(lam+3/8)

+          a.mn[j] <- mean(A)
+          a.vr[j] <- var(A)
+          a.sk[j] <- skewness(A)
+          a.ku[j] <- kurtosis(A)
+          }

+   d <- data.frame(f.mn, f.vr, f.sk, f.ku,
+                   a.mn, a.vr, a.sk, a.ku)
+   d
+ }
```

Note, the skewness and kurtosis functions require the e1071 package. After execution of the simulation and computation of the empirical mean values (over the simulations), for each of the transformed variates we obtain:

```
> simvalues <- code.hf1()
> round(apply(simvalues, 2, mean),2)
> f.mn   f.vr  f.sk   f.ku   a.mn   a.vr  a.sk   a.ku
  0.00   0.99  0.00  -0.63  -0.15   0.92 -0.11  -0.41
```

Several aspects are apparent from these simulation results. First, the bias of the Fisz-transformed variables is zero (to two decimal places) but that of Anscombe is -0.15. Second, the variance of the Fisz values is 0.99, very close to 1, whereas with Anscombe it is 0.92. The Fisz values are, on average, perhaps less skewed than Anscombe. The kurtosis values for Fisz are not as good as for the Anscombe transform (note these are corrected kurtosis values and have had three subtracted from them already). A slight word of warning is required here as Fisz requires two sets of Poisson random variates whereas Anscombe's requires only one. So the comparison here is somewhat naive.

So far, however, we have reminded the reader of the Fisz transform. However, this is a transform that maps two numbers to one, which is not quite what we require of a variance stabilization technique for a sequence of n items.

6.3 Poisson Intensity Function Estimation

To stabilize variance for a single random variable X, the only real option is to apply a univariate function. The methods in the previous section suggest which function to use to do this to get a good, although usually approximate, variance stabilization.

However, now suppose we have a sequence of random variables $X_1, \ldots, X_n$, where each X_i is independently distributed and depends on some parameter, say, λ_i. In many problems we might assume that λ_i is, in actuality, some observation from a function $\lambda(t)$, where $\lambda_i = \lambda(i/n)$. Further, we might decide to impose certain mathematical conditions upon λ. Typically, we might assume λ has a certain degree of smoothness or is, for example, piecewise constant.

The problem still is to variance stabilize, but now we wish to do this with the whole sequence of X_i. We could stabilize each X_i individually without reference to any of the other X_j for $j \neq i$. For example, if X_i were Poisson distributed, then we can still apply Anscombe's transform individually for each X_i. However, there is obviously a wider variety of things that we could do with the *whole* sequence to stabilize variance. The next section develops a function that takes in the whole sequence of n random variables and produces n outputs that are variance stabilized.

6.4 The Haar–Fisz Transform for Poisson Data

The motivation for our multiscale variance stabilizing transform stems from Theorem 6.1 and the remarks following it. There we noticed that the numerator of the Fisz-transformed variates was the Haar mother wavelet coefficient, and the denominator was the Haar scaling function coefficient.

6.4.1 Fisz transform of wavelet coefficients at the finest scale.

For the moment let us consider the discrete Haar wavelet transform as described in Section 2.1.2. We shall also assume that $n = 2^J$, although this restriction can be overcome fairly easily. The Fisz operation given in (6.14) can be written, for the first location index $k = 1$, as $f_{J-1,1} = d_{J-1,1}/c_{J-1,1}$. If the X_i are Poisson with intensity λ_i, then we know that $f_{J-1,1}$ is approximately normally distributed with a particular mean and variance of $1/\sqrt{2}$ (note not unit variance because of the different scaling of the Haar coefficients given in Section 2.1.2 and those in the Fisz formula in (6.14)). We can go further and repeat the formation of this kind of ratio for the different location indices $k = 1, \ldots, n/2$, i.e. form $f_{J-1,k} = d_{J-1,k}/c_{J-1,k}$ for those indices. Each of the different $f_{J-1,k}$ are formed from different X_i, and so they are all independent.

6.4.2 Fisz transform of wavelet coefficients at coarser scales.

At the next coarsest scale $(J - 2)$ we also have father coefficients, e.g.

$$c_{J-2,1} = (X_1 + X_2 + X_3 + X_4)/2, \tag{6.15}$$

and mother coefficients

$$d_{J-2,1} = (X_1 + X_2 - X_3 - X_4)/2. \qquad (6.16)$$

These can be written $Y_1 + Y_2$ and $Y_1 - Y_2$, where $Y_1 = X_1 + X_2$ and $Y_2 = X_1 - X_2$. The important thing to note is that because of the properties of Poisson random variables both Y_1 and Y_2 are distributed as Poisson: $Y_1 \sim \text{Pois}(\lambda_1 + \lambda_2)$ and $Y_2 \sim \text{Pois}(\lambda_3 + \lambda_4)$.

Hence, if we form $f_{J-2,1} = d_{J-2,1}/c_{J-2,1}$ it is again the case that the numerator is the difference of two Poisson random variables, and the denominator is the sum. So, again, by Fisz's theorem, the $f_{J-2,1}$ is approximately normal with some mean and variance of $1/2$. Indeed, the same thing can be made to happen at every scale and location. The wavelet and father wavelet coefficients for each scale-location pair are obtained from the father wavelets at the previous scale (this is just the Haar wavelet algorithm), which are themselves merely sums of Poisson random variables (except at the finest scale, where the component random variables are the Poisson data values themselves). Then, when we form $f_{j,k}$, we always form a Fisz ratio just as in (6.14).

6.4.3 Means of Haar–Fisz coefficients

If we form ratios for *all* wavelet coefficients, we end up with a collection $f_{j,k}$ of random variables that are approximately Gaussian with unit variance. However, what are the means? Again, we turn to the finest scale and the first location ($j = J - 1$, $k = 1$). The Fisz coefficient here is precisely that in (6.14). If $\lambda_1 = \lambda_2$, then Theorem 6.1 says that $f_{J-1,1}$ has zero mean. If the two underlying intensities are unequal, then the mean of $f_{J-1,1}$ would be non-zero and the non-zero mean would be larger the further apart λ_1 and λ_2 were. What this means at the finest scale is that if λ_i is constant for a set of successive i, then those finest-scale Fisz coefficients reliant on those X_i would all be approximately scaled $N(0,1)$ random variables. As soon as one of the coefficients intersects a change in value in the λ_i intensity sequence, then the mean of any 'overlapping' Fisz coefficient will not be zero.

Practically speaking this is very convenient as it means that if we observe a large Fisz coefficient for some $f_{J-1,k}$, we know that with a high probability the underlying intensities of the random variables immediately to the left and right of k are likely to be different. Moreover, in this situation we have the advantage of being fairly sure on what constitutes a 'large' coefficient as we know that, approximately, the Fisz coefficient is normally distributed with a *known* variance. Contrast this to the usual situation in wavelet shrinkage where the variance has to be estimated.

6.4.4 Multiscale nature of Haar–Fisz

Finally, note that we form Fisz coefficients at all scales, not just the finest one. Here, a large Fisz coefficient (again simply judged against a suitably scaled $N(0, 1)$ variate) corresponds, with high probability, to a large difference between the intensities of the component random variables. Here, at scale $J - j$, the component random variables of the Fisz coefficient $f_{J-j,k}$ are $Y_1 = c_{J-j+1,k}$, $Y_2 = c_{J-j+1,k+1}$, which correspond to two consecutive sets of X_i of length 2^j. For example, if $j = 2$, then $c_{J-1,1} = 2^{-1/2}(X_1 + X_2)$ and $c_{J-1,2} = 2^{-1/2}(X_3 + X_4)$ as above. So the Fisz coefficient detects differences between the mean intensity over the indices $1, 2$ and compares it to the mean intensity over $3, 4$. In other words, the Fisz coefficients at scale $J - 2$ 'scan' the Poisson signal for large differences averaged over scale two; those at scale $J - 3$ 'scan' over differences averaged over scale four, and so on.

Hence, the complete set of Fisz coefficients contains information about changes in the underlying intensity at all scales and all locations within the original signal (*all* meaning 'all possibilities afforded by the dyadic wavelet nature').

6.4.5 Inverse Haar transform

To summarize, we now take all the Fisz coefficients in the wavelet domain and apply the inverse Haar wavelet transform. The Fisz coefficients, $f_{j,k}$, are all approximately Gaussian, are also approximately uncorrelated, see Fryzlewicz and Nason (2004), and have a particular mean vector. If the true Poisson intensity sequence is piecewise constant, then the mean vector of the $f_{j,k}$ is zero except near the jumps in the Poisson intensity sequence (and conceptually sparse if there are a 'not too large' number of jumps). If we assume a smooth intensity, then the means will be non-zero but generally small, i.e., the usual nice situation encountered in wavelet shrinkage.

At this stage, it would be possible to perform standard wavelet threshold-ing on the *Fisz* coefficients. Everything is nicely set up! The Fisz coefficients are now in a simple Gaussian signal-plus-noise model, and even the noise is (approximately) known. Hence, we have replaced a 'hard' problem, which is to estimate the mean where the variance is known to vary, with a very simple wavelet shrinkage set-up. Indeed, much Haar–Fisz intensity estimation does just this. It performs wavelet shrinkage on the Fisz coefficients, then undoes the Fisz normalization, using the known scaling function coefficients, and then inverts the Haar transform.

The other possibility at this stage is to apply the inverse Haar transform to the Fisz coefficients. This results in a sequence with a Gaussian signal-plus-noise structure with constant variance. This result is known as the Haar–Fisz transform of the Poisson data and is the variance-stabilized form of $X_1, \ldots, X_n$.

6.4.6 Formulae for Haar–Fisz transform

The following formulae are taken from Fryzlewicz and Nason (2004) but with the notation altered for consistency with this book.

The Haar–Fisz transform is computed as follows.

Forward Haar: Define $c^J = (X_1, \ldots, X_n)$ and perform the forward Haar transform by

$$c_k^{j-1} = (c_{2k-1}^j + c_{2k}^j)/2 \tag{6.17}$$

and

$$d_k^{j-1} = (c_{2k-1}^j - c_{2k}^j)/2, \tag{6.18}$$

and define c^j and d^j to be the vectors of c_k^j and d_k^j for all $k = 1, \ldots, 2^{j-1}$.

Fisz transform: Define f^j to be the vector of f_k^j to be defined by

$$f_k^j = \begin{cases} 0 & \text{if } c_k^j = 0, \\ d_k^j (c_k^j)^{-1/2} & \text{otherwise,} \end{cases} \tag{6.19}$$

and replace all d_k^j by f_k^j.

Inverse Haar: Apply the inverse Haar transform to the c_0^0 and modified $d_k^j = f_k^j$, i.e., compute

$$c_{2k-1}^j = c_k^{j-1} + f_k^{j-1} \tag{6.20}$$

and

$$c_{2k}^j = c_k^{j-1} - f_k^{j-1}, \tag{6.21}$$

for $k = 1, \ldots, 2^{j-1}$ and j going from 1 to J.

Call the final c_k^J vector, obtained from the Fisz–modified coefficients, u_k for $k = 1, \ldots, 2^J = n$.

The whole operator from vector X to u is the Haar–Fisz transform (for Poisson data) and can be denoted by $\mathcal{F}X = u$, where $\mathcal{F}$ is the Haar–Fisz operator. Clearly $\mathcal{F} : \mathbb{R}^n \to \mathbb{R}^n$.

Fryzlewicz and Nason (2004) point out the main differences between the Haar wavelet transform and the Haar–Fisz transform are simple and single mathematical operations (a division and a square root). The computational complexity of the two algorithms in terms of orders of magnitude is the same. Hence, the Haar–Fisz transform is a fast algorithm. The other point to note with the transform in this section is that the forward Haar wavelet filters are $(1/2, -1/2)$, compared to $(1/\sqrt{2}, -1/\sqrt{2})$ of the Haar transform given in Section 2.1.2. So this transform is orthogonal but not orthonormal.

6.4.7 The inverse Haar–Fisz transform

What about the inverse Haar–Fisz transform, $\mathcal{F}^{-1}$? This can simply be achieved by reversing the above steps: take the Haar wavelet transform of u, remultiply the f_k^j coefficients by $(c_k^j)^{-1/2}$, and then perform the inverse Haar wavelet transform. Again, the inverse Haar–Fisz transform has the same computational complexity as the inverse Haar transform.

Fryzlewicz and Nason (2004) give the full formula for u in terms of X for the simple case $n = 8$. The first formula in this sequence, which gives the first Haar–Fisz-transformed value u_1, is given by

$$u_1 = \frac{\sum_{i=1}^{8} X_i}{8} + \frac{\sum_{i=1}^{4} X_i - \sum_{i=5}^{8} X_i}{2\sqrt{2}\sqrt{\sum_{i=1}^{8} X_i}} + \frac{X_1 + X_2 - (X_3 + X_4)}{2\sqrt{\sum_{i=1}^{4} X_i}} + \frac{X_1 - X_2}{\sqrt{2}\sqrt{X_1 + X_2}}.$$

$$(6.22)$$

In this small example, the structure of the Haar–Fisz transform can be clearly seen. The last three terms of the right-hand side of (6.22) are all examples of the Fisz transform. The first of these applies it to $X_1 + X_2 + X_3 + X_4$ and $X_5 + X_6 + X_7 + X_8$, the next to $X_1 + X_2$ and $X_3 + X_4$, and the last to X_1 and X_2. The next formula, for u_2, is the same is (6.22), except the last term is

$$\frac{X_3 - X_4}{\sqrt{2}\sqrt{X_3 + X_4}}.$$

$$(6.23)$$

Formulae (6.22) and (6.23) begin to show the multiscale nature of the variance stabilization. In going from u_1 to u_2 the large-scale contributions (the second and third terms of (6.22)) do not change, but the fine scale component stabilizing $\{1, 2\}$ changes to $\{3, 4\}$, if we moved from u_1 to much higher u_i, then the larger scale contributions would also change. So one should really imagine $u_1, \ldots, u_n$ laid out in sequence and for a given u_i an expanding inverted pyramid of individual stabilization terms contributing to that value. In particular, if there were a change in the underlying intensity around the location of X_i (and hence u_i), then this gets picked up by all the stabilization terms 'above' u_i. If the underlying intensity is piecewise constant up to 2^j terms either side of u_i, then the j terms 'above' u_i will all have zero mean (but larger-scale ones above those a non-zero mean). A depiction of the inverted pyramid for a particular u_i value for $n = 8$ is shown in Figure 6.2.

$$\text{Scale 8} \quad \left| \frac{X_1 + X_2 + X_3 + X_4 - (X_5 + X_6 + X_7 + X_8)}{2\sqrt{2}\sqrt{X_1 + X_2 + X_3 + X_4 + X_5 + X_6 + X_7 + X_8}} \right|$$
$$\text{Scale 4} \quad \left| -\frac{X_1 + X_2 - (X_3 + X_4)}{2\sqrt{X_1 + X_2 + X_3 + X_4}} \right|$$
$$\text{Scale 2} \quad \left| -\frac{X_3 - X_4}{\sqrt{2}\sqrt{X_3 + X_4}} \right|$$
$$u_1 \ u_2 \ u_3 \qquad u_4 \qquad u_5 \ u_6 \ u_7 \qquad u_8$$

Fig. 6.2. Inverted pyramid above u_3 consisting of all Haar–Fisz terms not including sample mean. The *vertical bars* are only to show horizontal extent and not absolute value.

6.4.8 The Haar–Fisz transform in WaveThresh

As an example we first set up a simple 'underlying intensity' function as shown in the top left of Figure 6.3. The intensity actually is

$$\lambda_i = \begin{cases} 1, & i = 1, \ldots, 256 \\ 10, & i = 257, \ldots, 512. \end{cases} \qquad (6.24)$$

A realization from this $\{\lambda_i\}_{i=1}^{512}$ sequence is depicted in the top right of Figure 6.3. Clearly, as expected, the variance on the high-intensity part is greater than on the low-intensity part (of course, it is ten times as great). The

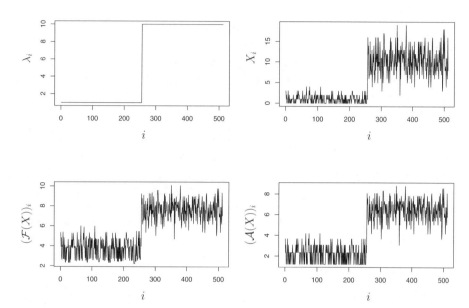

Fig. 6.3. *Top left*: example intensity function, λ_i. *Top right*: a simulated Poisson sequence $X_i \sim \text{Pois}(\lambda_i)$. *Bottom left*: Haar–Fisz transform of X_i. *Bottom right*: Anscombe transform of X_i. Produced by `f.hf2()`.

bottom row shows the results after stabilizing with the Haar–Fisz transform (left) and the Anscombe transform (right). The WaveThresh function `hft` computes the Haar–Fisz transform for Poisson data. The code to produce these figures is reproduced next.

```
#
# Arrange 2x2 plot
#
> op <- par(mfrow=c(2,2))

#
# Initialize true intensity vector
#
> lambda <- c(rep(1,256), rep(10,256))

#
```

```
# Calculate length of this vector
#
> n <- length(lambda)


#
# Draw a realization, one Poisson X for each i
#
> X <- rpois(n, lambda=lambda)


#
# Plot the intensity and the realization
#
> plot(1:n, lambda, type="l", xlab="i", ylab="lambdai")
> plot(1:n, X, type="l", xlab="i", ylab="Xi")


#
# Compute Haar-Fisz and Anscombe transforms
#
> XHF <- hft(X)
> XAn <- 2*sqrt(X+3/8)


#
# Plot stabilized sequences
#
> plot(1:n, XHF, type="l", xlab="i", ylab="FXi")
> plot(1:n, XAn, type="l", xlab="i", ylab="Anscombei")
```

We also arrange for our code to compute the variance on the low- and high-intensity parts of both stabilization results.

```
> vhf1 <- var(XHF[1:256])
> vhf2 <- var(XHF[257:512])
> va1 <- var(XAn[1:256])
> va2 <- var(XAn[257:512])

> cat("Variance of HF on first section is ",
+     vhf1, "\n")
> cat("Variance of HF on second section is ",
+     vhf2, "\n")

> cat("Variance of Anscombe on first section is ",
+     va1, "\n")
> cat("Variance of Anscome on second section is ",
+     va2, "\n")
```

The output of these variance calculations for Figure 6.3 is

```
Variance of HF on first section is  0.9244394
Variance of HF on second section is  1.076124
Variance of Anscombe on first section is  0.6967095
Variance of Anscome on second section is  1.088012
```

The results of the variance computation show that the variances for Haar–Fisz are closer to one and, probably more to the point, the variances for the two segments for Haar–Fisz are closer than for Anscombe. Many more simulations of this kind, as well as empirical evidence for the improved Gaussianization of Haar–Fisz over Anscombe, can be found in Fryzlewicz and Nason (2004).

For another example, we derive an intensity sequence from the Donoho and Johnstone (1994b) Blocks function. The commands to obtain this and then modify it so that the intensities ≥ 0 are:

```
> lambda <- DJ.EX(n=512)$blocks
> lambda <- (lambda - min(lambda) + 1)/8
```

These replace the assignment of `lambda` given in the previous code segment.

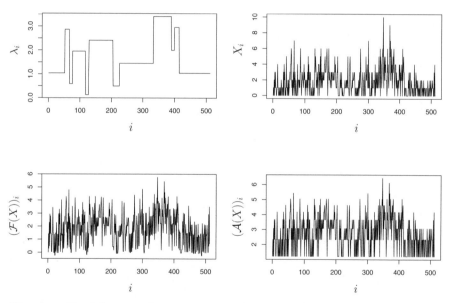

Fig. 6.4. *Top left*: example intensity function, λ_i based on Blocks function. *Top right*: a simulated Poisson sequence $X_i \sim \text{Pois}(\lambda_i)$. *Bottom left*: Haar–Fisz transform of X_i. *Bottom right*: Anscombe transform of X_i. Produced by `f.hf3()`.

The rest of the function that performs the Haar–Fisz and Anscombe transforms is the same. Figure 6.4 shows the realization and the result of both transforms. This time it is a little less obvious that the variance has been stabilized effectively (although it has), but clearly the Haar–Fisz has a more continuous appearance and less discrete nature than that produced by the Anscombe transform. The next section considers denoising both the sequences in the bottom row of Figure 6.4 and inverting the transforms to obtain an estimate. The two examples stated here are actually piecewise constant. However, Fryzlewicz and Nason (2004) demonstrate that the nice properties due to Haar–Fisz extend to smoother underlying intensity functions as well.

6.4.9 Denoising and intensity estimation

With the Haar–Fisz transform we know that the transformed sequence has an approximate signal+noise representation with Gaussian noise and stabilized variance. Hence, we can use any appropriate denoiser suitable for Gaussian noise. We recommend using wavelet shrinkage methods as they have excellent all-round performance but also perform extremely well on signals with discontinuities and other inhomogeneities.

Here, we choose to denoise both the Haar–Fisz and Anscombe transformed intensities that were simulated from the Blocks signal above. We use Haar wavelets but generally others could be used. We choose to use the EbayesThresh package described above in Section 3.10.4. The code to do this is

```
> XHFwd <- wd(XHF, filter.number=1, family="DaubExPhase")
> XAnwd <- wd(XAn, filter.number=1, family="DaubExPhase")

> XHFwdT <- ebayesthresh.wavelet(XHFwd)
> XAnwdT <- ebayesthresh.wavelet(XAnwd)

> XHFdn <- wr(XHFwdT)
> XAndn <- wr(XAnwdT)

> plot(1:n, XHFdn, type="l", xlab="i.")
> plot(1:n, XAndn, type="l", xlab="i.")

> XHFest <- hft.inv(XHFdn)
> XAnest <- ((XAndn)/2)^2 - 3/8

> plot(1:n, XHFest, type="l", xlab="i.")
> lines(1:n, lambda, lty=2)
> errHF <- sqrt(sum((XHFest-lambda)^2))
> plot(1:n, XAnest, type="l", xlab="i.")
> lines(1:n, lambda, lty=2)
> errAn <- sqrt(sum((XAnest-lambda)^2))

> cat("HF error is ", errHF, "\n")
> cat("An error is ", errAn, "\n")
```

Denoised versions of the intensities are stored in the `XHFdn` and `XAndf` for the Haar–Fisz and Anscombe transformed realization respectively. Then the inverse transformation is applied resulting in estimates `XHFest` and `XAnest`. Note that the last lines of this code compute the error between the two estimates and the truth, λ. The results of the denoising and estimation are shown in Figure 6.5. Clearly the Haar–Fisz-based estimate is considerably

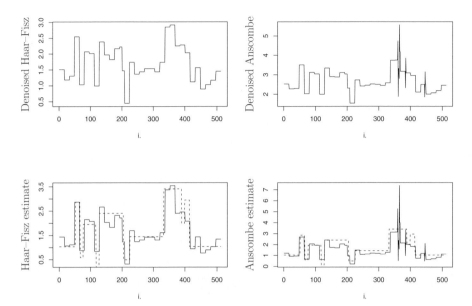

Fig. 6.5. *Top left*: denoised Haar–Fisz transformed intensities. *Top right*: denoised Anscombe transformed intensities *Bottom left*: Haar–Fisz estimate of λ_i. *Bottom right*: Anscombe estimate of λ_i. *Dotted line* in bottom row is original λ_i sequence. Produced by `f.hf4()`.

better. Indeed, the root mean square error as calculated in the code is better for Haar–Fisz too: it is 10.7 versus 14.0 for Anscombe.

6.4.10 Cycle spinning

Fryzlewicz and Nason (2004) also use TI denoising as described in Section 3.12.1. This improves the denoising. However, Fryzlewicz and Nason (2004) also use what they term as 'external' TI denoising, or more precisely 'external' cycle spinning as only a portion of the full cycle spins are used (or indeed seem to be required). The external cycle spinning works by generating full estimates (e.g. as in the previous section, potentially also with 'internal' TI denoising), and then rotating the input vector and repeating the denoising. This repeat rotation estimate is performed 10 to 50 times and the

average estimate returned. This kind of external cycle spinning is not worth-
while using Anscombe because the transform is diagonal and the Anscombe
operator commutes with the shift operator, whereas the same is not true for
the Haar–Fisz operator. Fryzlewicz and Nason (2004) show that using external cycle spinning improves the Haar–Fisz performance even further and for
many problems can be considered to be extremely competitive with existing
methods, see e.g., Kolaczyk (1999a) and Timmermann and Nowak (1999).

6.5 Data-driven Haar–Fisz

Almost this entire section so far has been concerned with variance stabilization
when the random variable of interest is distributed as Poisson. It is a natural
question to speculate about whether one might attempt something similar for
variables with distributions other than Poisson. Indeed, it is already known
that the log transformation can be useful in certain situations, e.g. such as
for handling χ^2 random variables that occur in areas such as periodogram
estimation in time series analysis, see Brockwell and Davis (1991), for example.

More common is the situation where one does not know the distribution of
the random variable of interest, or, at best, one only knows it approximately.
For this situation Fryzlewicz and Delouille (2005) and Fryzlewicz et al. (2007)
have developed the *data-driven* Haar–Fisz transform (DDHFT), which is
designed to stabilize under the following set-up.

Let $\mathbf{X} = (X_i)_{i=1}^n$ denote a generic input vector to the DDHFT. The
following list specifies the generic distributional properties of $\mathbf{X}$.

1. The length n of $\mathbf{X}$ must be a power of two. We denote $J = \log_2(n)$.
2. $(X_i)_{i=1}^n$ must be a sequence of independent, nonnegative random variables
   with finite positive means $\mu_i = \mathbb{E}(X_i) > 0$ and finite positive variances
   $\sigma_i^2 = \mathrm{Var}(X_i) > 0$.
3. The variance σ_i^2 must be a non-decreasing function of the mean μ_i:

$$\sigma_i^2 = h(\mu_i), \tag{6.25}$$

where the function h is independent of i.

Clearly, Poisson distributed variates satisfy these conditions with $h(x) = x$.
The χ^2 distribution is another example. A χ_ν^2 random variable on ν degrees
of freedom has mean ν and variance of 2ν, and hence here $h(x) = 2x$. There
are several other examples.

6.5.1 h known

For the Poisson case Formula (6.19) achieves a variance stabilized coefficient,
f_k^j, by dividing the wavelet coefficient d_k^j by an estimate of its standard
deviation—in this case $(c_k^j)^{1/2}$. However, here the mean-variance relationship

is encoded by h, and so this needs to play a part in the variance stabilization. For simplicity let us consider the finest-scale Haar wavelet coefficient d_1^{J-1} as in Fryzlewicz et al. (2007). From (6.18) we know that $d_1^{J-1} = (X_1 - X_2)/2$. Hence, under the assumption that X_1, X_2 are identically distributed,

$$\text{var}(d_1^{J-1}) = \{\text{var}(X_1) + \text{var}(X_2)\}/4 = \sigma_1^2/2, \tag{6.26}$$

which gives

$$2^{-1/2}\text{var}(d_1^{J-1})^{1/2} = \sigma_1, \tag{6.27}$$

and, of course, $\sigma_1 = h^{1/2}(\mu_1)$. In practice, we do not know μ_1 but we can estimate it locally by $c_1^{J-1} = (X_1 + X_2)/2$. Hence, the coefficient

$$f_1^{J-1} = \frac{d_1^{J-1}}{h^{1/2}(c_1^{J-1})}, \tag{6.28}$$

again with the convention that $0/0 = 0$, will be an approximately variance-stabilized coefficient.

We can repeat the above operation with other d_k^j and c_k^j in a similar fashion with successful stabilization occurring whenever the X_i that comprise the coefficients have the same distribution. If with d_1^{J-1} the components X_1 and X_2 do not have the same distribution, then the mean of d_1^{J-1} is no longer zero and then the transformed coefficient will deviate from a normal mean zero distribution and hence carry *signal*, information that some transition between 'intensities' has occurred between X_1 and X_2. This information is retained by subsequent thresholding, just as described above in Sections 6.4.3 and 6.4.9.

Incidentally, if our **X** were Poisson distributed, then with $h(x) = x$ Formula (6.28) reduces to (6.19).

6.5.2 h unknown

In practice, h is often unknown and has to be estimated from the data. Naturally, there are many ways in which h could be estimated. Fryzlewicz et al. (2007) suggest the following procedure. Since $\sigma_i^2 = h(\mu_i)$, it would be sensible to compute the empirical variances of $X_1, X_2, \ldots$ at locations $\mu_1, \mu_2, \ldots$ and then to smooth these to obtain an estimate of h. How can we estimate the empirical variances? Well, with a suitable smoothness assumption on the evolution of the σ_i^2 we can estimate σ_i^2 by

$$\hat{\sigma}_i^2 = \frac{(X_i - X_{i+1})^2}{2}. \tag{6.29}$$

On a very smooth (piecewise constant) stretch where $\sigma_i^2 = \sigma_{i+1}^2$ it turns out that $\mathbb{E}(\hat{\sigma}_i^2) = \sigma_i^2$. This discussion motivates the following 'regression' model:

$$\hat{\sigma}_i^2 = h(\mu_i) + \epsilon_i, \tag{6.30}$$

where $\epsilon_i = \hat{\sigma}_i^2 - \sigma_i^2$ and for smooth intensities we would expect that $\mathbb{E}(\epsilon_i) \approx 0$, and for piecewise constant intensities we would expect the expectation to be exactly zero.

Further, in practice, we do not know the μ_i, so Fryzlewicz et al. (2007) estimate that too by

$$\hat{\mu}_i = \frac{X_i + X_{i+1}}{2},$$

and hence h is estimated via the finest-scale wavelet coefficients. There are other possible variants on this scheme including those that use coarser-scale coefficients. In the data-driven model (6.30) above, Fryzlewicz et al. (2007) assume that h is non-decreasing and hence use the monotonic regression method described in Johnstone and Silverman (2005a) (and based on software written by Bernard Silverman), which produces a non-decreasing solution (and the mathematics necessarily dictates that it is also piecewise constant solution).

The *data-driven* HFT proceeds as in the previous section but with $\hat{h}$ replacing h.

Figure 6.6 shows two test examples from Fryzlewicz et al. (2007) where *we* know the true distribution of the data but, of course, the DDHFT does not have this information. The top row of Figure 6.6 shows realizations from both a Poisson sequence and a set of scaled independent χ_2^2 random variables where, in both cases, the underlying intensity/scaling is the Blocks function (scaled to have minimum/maximum intensities of 3 and 25 respectively). The second row shows the estimate of the mean-variance function, h as a solid line, and the true underlying h in each case is shown. For Poisson the truth is, of course, $h(\mu) = \mu$ and for the χ_2^2 it is $h(\mu) = \mu^2$. In both cases, the estimate $\hat{h}$ of h does look good (although the estimate could benefit from confidence/credible intervals). The third row shows the DDHFT of each respective sequence shown in the top row. The bottom row shows the classical stabilization transforms for each of these types of data: the Anscombe transform of the top row (for Poisson) and the log-transformed data (for χ^2).

Fryzlewicz et al. (2007) note that visually, there is not a lot to choose between the DDHFT and Anscombe transform for the Poisson data. However, remember that the Anscombe transform assumes that the data are Poisson (and hence 'knows' h) whereas the DDHFT does not rely on this assumption and has estimated h. For the χ^2 data Fryzlewicz et al. (2007) argue that the DDHFT-transformed version is more symmetric and that the shape of the underlying signal is clearer—again, remarkable given that the DDHFT has to estimate h.

Further simulations by Fryzlewicz et al. (2007) show that overall performance of the DDHFT is hardly worse than that of the HFT (i.e. where h is assumed known) and occasionally it is better.

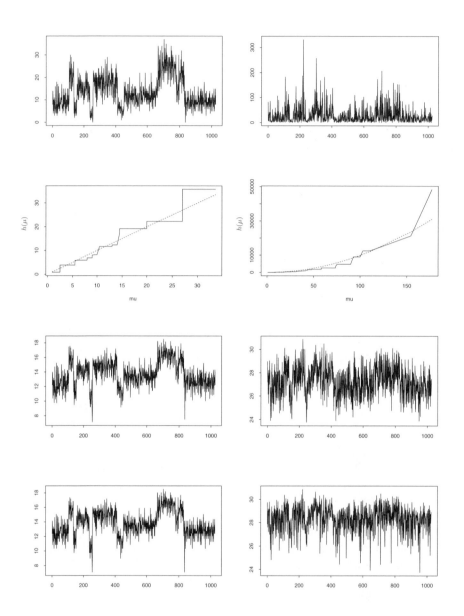

Fig. 6.6. *Left-hand column* corresponds to the Poisson case and *right-hand column* to the χ^2 case. *Top row*: Blocks function 'contaminated' with selected noise. *Second row*: the true variance function h (*dotted line*) and its DDHFT estimate (*continuous line*). *Third row*: DDHFT of both signals from *top row* using the DDHFT estimated h. *Bottom row*: Poisson data stabilized by the Anscombe transform (*left*), and the χ^2_2 data stabilized via the log transform (*right*). Figure 3 from Fryzlewicz et al. (2007). Reproduced with permission. Produced by Piotr Fryzlewicz

6.5.3 (Worked) Example 1: airline passenger data

Figure 6.7 shows the famous airline passenger data taken from Box et al. (1994) (part of R in the AirPassenger data set). This well-known data set exhibits a number of clear and interesting features as follows: (i) the local mean of the series increases over time, (ii) there is a clear annual seasonal effect, and (iii) most relevant for our discussion, the variability of the series increases over time. We can apply our data-driven Haar–Fisz algorithm in R

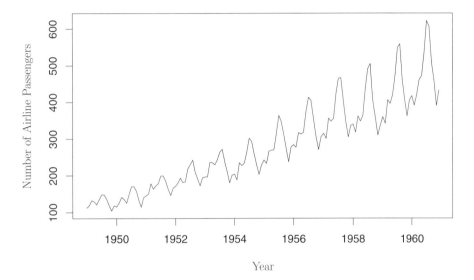

Fig. 6.7. Monthly number of airline passengers (in thousands of passengers per month). Produced by f.hf6().

by first loading the DDHFm package and then applying the ddhft.np.2 function to perform the DDHF as follows:

```
#
# Only analyze the first 128 observations
#
> AP128 <- AirPassengers[1:128]

#
# Convert those obs into a monthly time series
#
> AP128 <- ts(AP128, start=c(1949,1), frequency=12)

#
# Perform the DDHF
#
```

```
> APhft <- ddhft.np.2(AP128)

#
# Plot the mean-variance relationship
#
> plot(APhft$mu, sqrt(APhft$sigma2), xlab="MU", ylab="SIGMA")
> lines(APhft$mu, APhft$sigma)
> lines(APhft$mu, sqrt(APhft$mu), lty=2)
```

Due to construction of the fast Haar transform algorithm in ddhft.np.2 the function can only handle data that have dyadic length. In principle, the algorithm can handle data of any length but new code would need to be written to do this. To work round this we only analyze the first 128 observations of AirPassenger (another possibility would be to pad the series at either end with zeroes, or possibly a tapered extension). The estimated

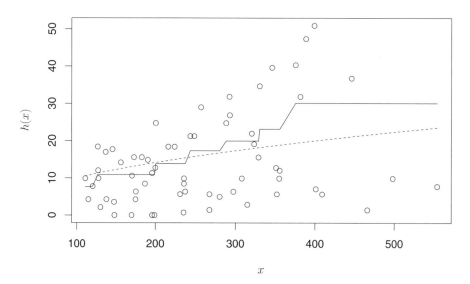

Fig. 6.8. *Circles*: estimated standard deviations, $\hat{\sigma}_i^2$, plotted against estimated means, $\hat{\mu}_i$. *Solid line*: monotonic regression line fitted to $(\hat{\mu}_i, \hat{\sigma}_i^2)$. *Dotted line*: square-root function. Produced by f.hf7().

mean-variance relationship $\hat{h}$ and the estimated local means and standard deviations from the above ddhft.np.2 analysis are shown in Figure 6.8. The dotted line in Figure 6.8 corresponds to the square-root function (the log function could also be drawn but follows the monotonic regression line much less closely—even allowing a scale factor). It is at least plausible that the mean-variance relationship is $h(\mu) \approx \mu$.

However, much care should be employed when attempting to interpret figures such as Figure 6.8. Figure 6.9 shows the DDHF-transformed time series as a solid line. The DDHF has stabilized the variance pretty well. The peaks at the end of the series are only slightly bigger than those at the beginning (compared to the large differences at the beginning and end of the original `AirPassenger` series shown in Figure 6.7). The square root function, which is clearly located *under* the DDHF-transformed series, has also stabilized well, but not as well as the DDHF. However, the log transform is extremely similar to the DDHF version for this data set. Hence, one might have assumed that,

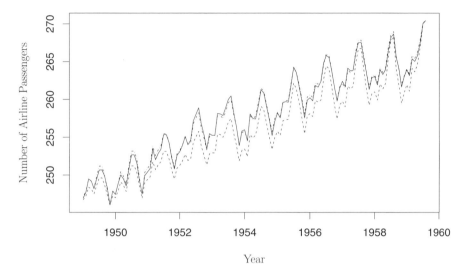

Fig. 6.9. *Solid line*: DDHF-transformed `AirPassenger` data. *Dashed line*: `AirPassenger` data subjected to the square-root transform. *Dotted line*: `AirPassenger` data subjected to the log transform. Both the square root and log transformed variables have been further subjected to independent linear scaling and shifting to match the range of the DDHF-transformed data. Produced by `f.hf8()`.

since Figure 6.8 seemed to indicate a square root mean-variance relationship there was a Poisson-like behaviour in the `AirPassenger` series. However, the actual mean-variance relationship is subtly different and, as shown by Figure 6.9, operates much like the log transform. The operation and result of the DDHF transform gives strong support to the use of the log transform in this situation. Remember that the DDHF placed no distributional assumption upon the data and had to estimate the mean-variance relationship. However, the relationship it found concurs with the operation of the log transform.

6.5.4 Example 2: GOES-8 X-ray data

Fryzlewicz et al. (2007) discuss a set of solar flux time series obtained from the X-ray sensor (XRS) from the GOES-8 satellite. The XRS provides background

X-ray information and warns when a solar flare has occurred. This information can be used to predict a solar-terrestrial disturbance on Earth which can disrupt communications and navigation systems, and even disable power grids. See Fryzlewicz et al. (2007) for further information and references.

Figure 6.10 (top) shows an example of an XRS data set consisting of $2^{14} = 16384$ observations recorded every 3 seconds from about 10am to midnight on 9th February 2001. The observations represent measurements for the whole Sun X-ray flux in the 0.1–0.8nm wavelength band. The series is noisy and, although the variance of the noise is not large, it does appear to depend on the mean intensity of the signal. It is of interest to be able to remove the noise, both for examination of the signal and to help any post-processing of the signal.

Figure 6.10 (bottom) shows the results of applying universal thresholding, as described in Section 3.5. Our application of universal thresholding is rather reckless because the method assumes that the variance of the noisy data is constant—which is not the case here. The 'denoised' signal at the bottom of Figure 3.5 is alternately very smooth and very noisy: this presumably corresponds to less and more noisy parts of the original signal. Fryzlewicz et al. (2007) try using the AVAS and ACE variance stabilization techniques of Breiman and Friedman (1985) and Tibshirani (1988) but report poor performance on this example; still the noise is evident in some areas but not others.

Fryzlewicz et al. (2007) combine the DDHF algorithm with a wavelet smoothing method on the XRS data. First, the variance function is estimated from the data. The top left plot of Figure 6.11 shows the pairs $(c_k^{J-1}, 2(d_k^{J-1})^2)$ plotted on a square-root scale. These pairs are the Haar father and twice the square of mother wavelet coefficients of X_t respectively. The top right plot in Figure 6.11 shows the monotonically non-decreasing estimated function $\hat{h}(\mu)$ plotted on the same scale. The top right plot exhibits a fairly sharp step around $\mu \approx 1.2 \times 10^{-6}$, which indicates (at least) two possible noise regimes: one with a lower variance for $\mu < 1.2 \times 10^{-6}$ and one for μ above this value. It is conjectured in Fryzlewicz et al. (2007) that this sharp change is maybe due to the electronics in the sensor responding to changes in intensity of the received signal. The variance-stabilized sequence obtained after applying the DDHFT to X_t is shown in the middle left plot of Figure 6.11, and the variance appears to be approximately constant across the figure (more so than that produced by AVAS and ACE). The residuals from fitting a standard universally thresholded wavelet estimator are shown in the middle right plot and confirm the reasonable constancy of the variance across the signal.

The bottom left plot in Figure 6.11 shows the result of applying the inverse DDHFT to the universally thresholded DDHF-transformed estimate. The quality of the estimate, compared to other methods, is apparently very good—certainly much better than when universal thresholding was applied without variance stabilization above, partially shown as the bottom plot in Figure 6.10 (and better than AVAS and ACE, see Fryzlewicz et al. (2007)).

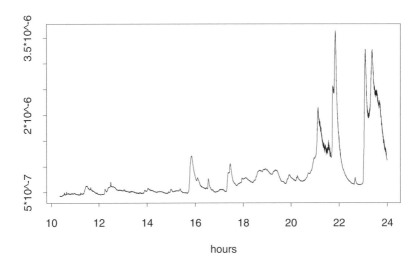

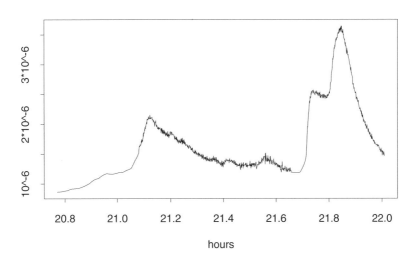

Fig. 6.10. *Top*: solar X-ray flux, X_t, recorded by GOES-8 XRS on February 9, 2001. Units are Wm^{-2}. *Bottom*: fragment of X_t denoised by the universal thresholding procedure. Reproduced with permission from Fryzlewicz et al. (2007).

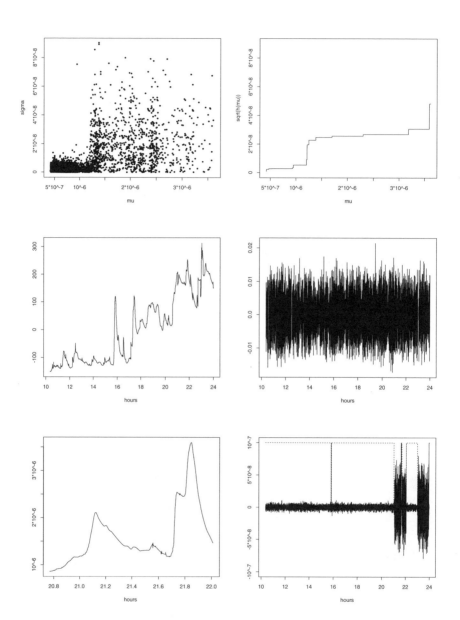

Fig. 6.11. DDHFT of XRS data, X_t. *Top row*: scaled moduli of the finest-scale Haar detail coefficients against the finest-scale Haar smooth coefficients (*left*), and the estimated variance function, $\hat{h}$ (right). *Middle row*: DDHFT of X_t, (*left*), and residuals when the DDHFT of X_t is denoised (*right*). *Bottom row*: final estimate of $\mathbb{E}(X_t)$ (*left*) and (*right*) its residuals (*solid line*), and indicator $10^{-7}\mathbb{I}(X_t < 1.2 \times 10^{-6})$ (*dotted line*). Units for *top plots* (x and y axes) and *bottom plots* (y axes) are Wm^{-2}. Reproduced with permission from Fryzlewicz et al. (2007).

The bottom right plot shows the residuals between the original series given in Figure 6.10 (top) and the DDHF-wavelet estimate shown in the bottom-left of Figure 6.11. The bottom right plot also simultaneously displays the places where the original series is less than (greater than) 1.2×10^{-6}. The residual plot definitely indicates that, approximately, when $X_t < 1.2 \times 10^{-6}$, the variance of the residuals (and original signal) is very low, and then when $X_t > 1.2 \times 10^{-6}$, the variance is very large. This is a further, strong, indication that there are two different variance regimes in this signal. The reader should also remember that all this is achieved under the mild assumptions in the introductory part of this section around Formula (6.25).

6.6 Discussion

The Haar–Fisz transformations above are part of a body of work that is beginning to reappraise the value of variance stabilization in statistics. There are many possible variations and extensions of the work above. It should be emphasized that Haar–Fisz is but one component of this new collection of ideas, which is loosely based around a fusion of a multiscale (or basis transform) and existing variance stabilization. For example, the multiscale Poisson intensity estimation procedures due to Kolaczyk (1999a) and Timmermann and Nowak (1999) predate the Haar–Fisz ideas above. Fadili et al. (2003) consider 2D Haar–Fisz with a Bayesian denoiser. Jansen (2006) developed conditional variance stabilization, which extends the Haar–Fisz concept to more general wavelets and not just to Haar. Fryzlewicz and Nason (2006) propose a Haar–Fisz transform for χ^2 data in the context of spectrum estimation for locally stationary time series. Fryzlewicz et al. (2006) estimate the volatility of a simple piecewise stationary process using a Haar–Fisz method. Motakis et al. (2006) propose using a Haar–Fisz transform for blocked data, i.e., not a 'time series' sequence but where there is no particular ordering of the underlying intensity. Nunes (2006) proposed address variance stabilization for binomial count data in which the Fisz transform is replaced by a stabilizer based on the known parametric form of the count variance. Zhang et al. (2008) consider Poisson stabilization and estimation using an extended version of the Anscombe transformation applied to filtered data and then employed on images using ridgelets and curvelets. Nason and Bailey (2008) apply the data-driven Haar–Fisz transform to coalition personnel fatalities to better estimate the underlying intensity when compared to current extra-variable linear methods. Additionally, Bailey (2008) investigates a maximum likelihood approach to Haar–Fisz variable transformation and consider data with both positive and negative count values.

A

R Software for Wavelets and Statistics

Here is a list and a brief description of some important R packages related to/that make use of wavelets. Such packages are important as they enable ideas to be reproduced, checked, and modified by the scientific community. This is probably just as important as the advertisement of the ideas through scientific papers and books. This is the 'reproducible research' view expressed by Buckheit and Donoho (1995).

Making software packages freely available is an extremely valuable service to the community and so we choose to acclaim the author(s) of each package by displaying their names! The descriptions are extracted from each package description from CRAN. The dates refer to the latest updates on CRAN, not the original publication date. Please let me know of any omissions.

adlift performs adaptive lifting (generalized wavelet transform) for irregularly spaced data, see Nunes et al. (2006). Written by Matt Nunes and Marina Knight, University of Bristol, UK, 2006.

brainwaver computes a set of scale-dependent correlation matrices between sets of preprocessed functional magnetic resonance imaging data and then performs a 'small-world' analysis of these matrices, see Achard et al. (2006). Written by Sophie Achard, Brain Mapping Unit, University of Cambridge, UK, 2007.

CVThresh carries out level-dependent cross-validation for threshold selection in wavelet shrinkage, see Kim and Oh (2006). Written by Donghoh Kim, Hongik University and Hee-Seok Oh, Seoul National University, Korea, 2006.

DDHFm implements the data-driven Haar–Fisz transform as described in 6.5, see references therein. Written by Efthimios Motakis, Piotr Fryzlewicz and Guy Nason, University of Bristol, Bristol, UK.

EbayesThresh carries out Empirical Bayes thresholding as described in section 3.10.4, see papers listed there. Written by Bernard Silverman, Oxford University, UK, 2005.

nlt non-decimated lifting transform (generalized wavelet transform) for irregularly spaced data, see Knight and Nason (2008). Written by Marina Knight, University of Bristol, UK, 2008.

rwt The Rice Wavelet Toolbox wrapper. The Rice Wavelet Toolbox is a collection of Matlab files for 1D and 2D wavelet and filter bank design, analysis, and processing. Written by P. Roebuck, Rice University, Texas, USA, 2005.

SpherWave carries out the wavelet transform of functions on the sphere and related function estimation techniques. See Li (1999), Oh (1999), Oh and Li (2004). Written by Hee-Seok Oh, Seoul National University and Donghoh Kim, Hongik University, Korea, 2007.

unbalhaar computes the unbalanced Haar transform and related function estimation, see Fryzlewicz (2007). Written by Piotr Fryzlewicz, Bristol University, UK, 2006.

waved wavelet deconvolution of noisy signals, see Section 4.9, Raimondo and Stewart (2007). Written by Marc Raimondo and Michael Stewart, University of Sydney, Australia, 2007.

wavelets computes and plots discrete wavelet transform and maximal overlap discrete wavelet transforms. Written by Eric Aldrich, Duke University, North Carolina, USA, 2007.

waveslim computes 1D, 2D, and 3D wavelet transforms, packet transforms, maximal overlap transforms, dual-tree complex wavelet transforms, and much more! Based on methodology described in Percival and Walden (2000) and Gencay et al. (2001). Written by Brandon Whitcher, dude, Translational Medicine and Genetics, GlaxoSmithKline, Cambridge, UK, 2007.

wmtsa software to accompany the book Percival and Walden (2000). Written by William Constantine, Insightful Corporation, and Donald Percival, Applied Physics Laboratory, University of Washington, Seattle, USA, 2007.

B

Notation and Some Mathematical Concepts

This appendix gives some extra information, notation, and definitions on the more mathematical aspects underlying the concepts of this book. More details can be found in a number of elementary mathematical texts. In the wavelet context the book, Daubechies (1992) is precise and concise and Burrus et al. (1997) is a very readable introduction.

B.1 Notation and Concepts

B.1.1 Function spaces

In mathematical terms the book deals mostly with measurable functions f on $\mathbb{R}$ such that

$$\int_{-\infty}^{\infty} |f(x)|^2 \, dx < \infty. \tag{B.1}$$

The space of functions satisfying (B.1) is denoted $L^2(\mathbb{R})$. In engineering terms this means that the function, interpreted as a signal, has finite energy.

We let $\mathcal{C}$ the space of continuous functions (on $\mathbb{R}$), $\mathcal{C}^k$ the space of all continuous functions with continuous derivatives of orders up to, and including, k, and $\mathcal{C}^\infty$ the space of all continuous functions with continuous derivatives of all orders.

B.1.2 Support of a function

The *support* of a function, f (denoted supp f), is the complement of the largest open set E with the property that $x \in E \implies f(x) = 0$. It is the 'maximal set on which f is non-zero and inclusive'.

The function f is compactly supported if supp f is a compact set (closed and bounded).

B.1.3 Inner product, norms, and distance

Given two functions $f, g \in L^2(\mathbb{R})$ we can define the *inner product* of these by

$$< f, g >= \int_{\mathbb{R}} f(x)\overline{g(x)} \, dx, \tag{B.2}$$

where $\overline{g(x)}$ denotes the complex conjugate of $g(x)$. The inner product can be used to gauge the 'size' of a function by defining the L^2 norm $|| \cdot ||_2$ as follows:

$$||f||_2^2 =< f, f >, \tag{B.3}$$

which leads naturally to the notion of the 'distance' between two functions $f, g \in L^2(\mathbb{R})$ as $||f - g||_2$.

B.1.4 Orthogonality and orthonormality

Two functions $f, g \in L^2(\mathbb{R})$ are said to be *orthogonal* if and only if $< f, g >= 0$. Define the *Kronecker delta* by

$$\delta_{ij} = \begin{cases} 1 \text{ if } i = j \\ 0 \text{ if } i \neq j, \end{cases} \tag{B.4}$$

for integers i, j.

A set of functions $\{\phi_1, \ldots, \phi_n\}$ for some $n = 1, 2, \ldots$ forms an *orthonormal set* if and only if

$$< \phi_i, \phi_j >= \delta_{ij}. \tag{B.5}$$

B.1.5 Vector space sums

Let V, W be subspaces of $L^2(\mathbb{R})$. We define the following spaces. The vector space sum of V and W is defined as

$$V + W = \{f + g : f \in V, g \in W\}. \tag{B.6}$$

The subspaces V and W are said to be orthogonal and denoted $V \perp W$ if every function V is orthogonal to every function in W (i.e. $V \perp W$ if and only if $< f, g >= 0$ for all $f \in V, g \in W$). We further define

$$V \oplus W = \{f + g : f \in V, g \in W, V \perp W\}. \tag{B.7}$$

B.1.6 Fourier transform

The Fourier transform of a function $f \in L^2$ is given by

$$\hat{f}(\omega) = (2\pi)^{-1/2} \int_{-\infty}^{\infty} f(x) \exp(-i\omega x) \, dx. \tag{B.8}$$

The inverse Fourier transform of $\hat{f}(\omega)$ is given by

$$f(x) = (2\pi)^{-1/2} \int_{-\infty}^{\infty} \hat{f}(\omega) \exp(i\omega x) \, d\omega. \tag{B.9}$$

B.1.7 Fourier series

Statisticians are probably most familiar with Fourier series that occur routinely in time series analysis. Given a stationary time series $\{X_t\}$ with autocovariance function $\gamma(\tau)$ for integers τ, the spectral density function, $f(\omega)$, can be written as

$$f(\omega) = (2\pi)^{-1} \sum_{\tau=-\infty}^{\infty} \gamma(\tau) \exp(-i\omega\tau), \tag{B.10}$$

where $\omega \in (-\pi, \pi]$. The inverse Fourier relationship is

$$\gamma(\tau) = \int_{-\pi}^{\pi} f(\omega) \exp(i\omega\tau) \, d\omega, \tag{B.11}$$

for integer τ.

There are many other slightly different formulations of this. For example, Formulae (2.50) and (2.84) are both forms of (B.10) but with different scale factors.

B.1.8 Besov spaces

In this book much of what we do applies to functions that belong to more interesting function spaces. The natural home for wavelets are Besov spaces; an informal discussion appears in Section 3.10.3. The Besov sequence space norm for a function $f(x)$ is defined in terms of its wavelet coefficients $\{d_{j,k}\}$ and coarsest-scale coefficient $c_{0,0}$ by

$$||d||_{b_{p,q}^s} = |c_{0,0}| + \left\{ \sum_{j=0}^{\infty} 2^{js'q} \left(\sum_{k=0}^{2^j-1} |d_{j,k}|^p \right)^{q/p} \right\}^{1/q}, \tag{B.12}$$

if $1 \le q < \infty$ and

$$||d||_{b_{p,\infty}^s} = |c_{0,0}| + \sup_{j \ge 0} \left\{ 2^{js'} \left(\sum_{k=0}^{2^j-1} |d_{j,k}|^p \right)^{1/p} \right\}, \tag{B.13}$$

where $s' = s + \frac{1}{2} - 1/p$, see Abramovich et al. (1998). The sequence space norm is equivalent to a Besov function space norm on f, denoted $||f||_{B_{p,q}^s}$. Hence, one can test membership of a Besov space for a function by examining its wavelet coefficients (for a wavelet with enough smoothness).

B.1.9 Landau notation

We use Landau notation to express the computational efficiency of algorithms. Suppose $f(x)$, $g(x)$ are two functions. Then $f(x) = \mathcal{O}\{g(x)\}$ means that there exists x^* and a constant $K > 0$ such that for $x \geq x^*$ we have $|f(x)| \leq M|g(x)|$. For example, we use the notation in the following way. Suppose that we have an algorithm whose (positive) running time is $r(n)$, where the size of the problem is measured by n. Then we can say that the algorithm is $\mathcal{O}\{g(n)\}$ if there exists $K > 0$ such that for large enough n we have $r(n) \leq Kg(n)$. For example, multiplication of an $n \times 1$ vector by an $n \times n$ matrix is $\mathcal{O}(n^2)$.

C

Survival Function Code

This section contains the code for the remainder of the survival function estimation described in Section 4.8.

The function uk computes the U_k values as described in Antoniadis et al. (1999) as follows:

```
uk <- function(z, nbins=32){

anshfc <- hfc(z, nbins=nbins)
uk <- anshfc$bincounts/anshfc$del

ans <- list(uk=uk, bins=anshfc$bins, del=anshfc$del)

ans
}
```

The uk function is essentially a Δ-weighted version of hfc. The subf function below computes the subdensity estimator of $f^*(t)$ as follows:

```
subf <- function(z, nbins=32, filter.number=2, lev=1){

tmp <- uk(z, nbins=nbins)

zuk <- tmp$uk

yywd <- wd(zuk, filter.number=filter.number, bc="interval")

levs <- nlevels(yywd) - (1:lev)
yywdT <- nullevels(yywd, levelstonull=levs)

fhat <- wr(yywdT)

fhat[fhat < 0] <- 0
```

```
l <- list(fhat = fhat, bins=tmp$bins, del=tmp$del)
l
}
```

This function is very similar to the **hws** linear wavelet smooth in the main text with the addition that negative density values are prohibited and set to zero. The final, hazard estimation, function computes an estimate of the ratio $f^*(t)/L(t)$ by dividing the outputs of subf by Lest as follows:

```
hazest <- function(z, nbins=32, levN=1, levD=1,
    filter.number=8){

L <- Lest(z=z, nbins=nbins, lev=levD)

fsub <- subf(z=z, nbins=nbins, filter.number=filter.number,
    lev=levN)

haz <- fsub$fhat / L$L

ans <- list(haz=haz, bins=fsub$bins, del=fsub$del, L=L$L,
    fsub=fsub$fhat)
ans
}
```

References

Aalen, O. O. (1978) Nonparametric inference for a family of counting processes, *Ann. Statist.*, **6**, 701–726.

Abramovich, F. and Benjamini, Y. (1996) Adaptive thresholding of wavelet coefficients, *Comp. Stat. Data Anal.*, **22**, 351–361.

Abramovich, F. and Silverman, B. W. (1998) Wavelet decomposition approaches to statistical inverse problems, *Biometrika*, **85**, 115–129.

Abramovich, F., Sapatinas, T., and Silverman, B. W. (1998) Wavelet thresholding via a Bayesian approach, *J. R. Statist. Soc. B*, **60**, 725–749.

Abramovich, F., Bailey, T., and Sapatinas, T. (2000) Wavelet analysis and its statistical applications, *Statistician*, **49**, 1–29.

Abramovich, F., Besbeas, P., and Sapatinas, T. (2002) Empirical Bayes approach to block wavelet function estimation, *Comp. Stat. Data Anal.*, **39**, 435–451.

Abramovich, F., Benjamini, Y., Donoho, D. L., and Johnstone, I. M. (2006) Adapting to unknown sparsity by controlling the false discovery rate, *Ann. Statist.*, **34**, 584–653.

Abry, P. (1994) *Transformées en ondelettes — Analyses multirésolutions et signaux de pression en turbulance.*, Ph.D. thesis, Université Claude Bernard, Lyon, France.

Achard, S., Salvador, R., Whitcher, B., Suckling, J., and Bullmore, E. (2006) A resilient, low-frequency, small-world human brain functional network with highly connected association cortical hubs, *The Journal of Neuroscience*, **26**, 63–72.

Allan, D. W. (1966) Statistics of atomic frequency clocks, *Proceedings of the IEEE*, **54**, 221–230.

Anscombe, F. J. (1948) The transformation of Poisson, binomial and negative-binomial data., *Biometrika*, **35**, 246–254.

Antoniadis, A. (1997) Wavelets in statistics: a review, *J. Italian Stat. Soc.*, **6**, 97–130.

Antoniadis, A. (2007) Wavelet methods in statistics: some recent developments and their applications, *Statistics Surveys*, **1**, 16–55.

Antoniadis, A. and Fan, J. (2001) Regularization of wavelet approximations, *J. Am. Statist. Ass.*, **96**, 939–967.

Antoniadis, A. and Gijbels, I. (2002) Detecting abrubt changes by wavelet methods, *J. Nonparam. Statist.*, **14**, 7–29.

Antoniadis, A., Gregoire, G., and McKeague, I. M. (1994) Wavelet methods for curve estimation, *J. Am. Statist. Ass.*, **89**, 1340–1353.

Antoniadis, A., Gregoire, G., and Nason, G. P. (1999) Density and hazard rate estimation for right-censored data by using wavelet methods, *J. R. Statist. Soc.* B, **61**, 63–84.

Antoniadis, A., Gijbels, I., and MacGibbon, B. (2000) Non-parametric estimation for the location of a change-point in an otherwise smooth hazard function under random censoring, *Scand. J. Stat.*, **27**, 501–519.

Antoniadis, A., Bigot, J., and Sapatinas, T. (2001) Wavelet estimators in nonparametric regression: a comparative simulation study, *J. Statist. Soft.*, **6(6)**, 1–83.

Arias–Castro, E., Donoho, D. L., and Huo, X. (2005) Near-optimal detection of geometric objects by fast multiscale methods, *IEEE Trans. Inf. Th.*, **51**, 2402–2425.

Averkamp, R. and Houdré, C. (2003) Wavelet thresholding for non-necessarily Gaussian noise: idealism, *Ann. Statist.*, **31**, 110–151.

Bailey, D. (2008) *Data mining of Early Day Motions and multiscale variance stabilisation of count data.*, Ph.D. thesis, University of Bristol, U.K.

Barber, S. (2008) Personal communication.

Barber, S. and Nason, G. P. (2004) Real nonparametric regression using complex wavelets, *J. R. Statist. Soc.* B, **66**, 927–939.

Barber, S., Nason, G. P., and Silverman, B. W. (2002) Posterior probability intervals for wavelet thresholding, *J. R. Statist. Soc.* B, **64**, 189–205.

Barlett, M. S. (1936) The square root transformation in the analysis of variance, *J. R. Statist. Soc. Suppl.*, **3**, 68–78.

Barndorff-Nielsen, O. E. and Cox, D. R. (1989) *Asymptotic Techniques for Use in Statistics*, Chapman and Hall, London.

Benjamini, Y. and Hochberg, Y. (1995) Controlling the false discovery rate: a practical and powerful approach to multiple testing, *J. R. Statist. Soc.* B, **57**, 289–300.

Berkner, K. and Wells, R. O. (2002) Smoothness estimates for soft-threshold denoising via translation invariant wavelet transforms, *App. Comp. Harm. Anal.*, **12**, 1–24.

Bernardini, R. and Kovačević, J. (1996) Local orthogonal bases II: window design, *Multidimensional systems and signal processing*, **7**, 371–399.

Beylkin, G., Coifman, R. R., and Rohlkin, V. (1991) Fast wavelet transforms and numerical algorithms, *Comm. Pure Appl. Math.*, **44**, 141–183.

Bezandry, P. H., Bonney, G. E., and Gannoun, A. (2005) Consistent estimation of the density and hazard rate functions for censored data via the wavelet method., *Stat. Prob. Lett.*, **74**, 366–372.

Bouman, P., Dukic, V., and Meng, X.-L. (2005) A Bayesian multiresolution hazard model with application to an aids reporting delay study, *Statist. Sci.*, **15**, 325–327.

Bowman, A. W. and Azzalini, A. (1997) *Applied Smoothing Techniques for Data Analysis: The Kernel Approach with S-Plus illustrations*, Oxford University Press, Oxford.

Box, G. E. P., Jenkins, G. M., and Reinsel, G. C. (1994) *Time-Series Analysis, Forecasting and Control*, Holden-Day, San Francisco.

Breiman, L. and Friedman, J. H. (1985) Estimating optimal transformations for multiple regression and correlation, *J. Am. Statist. Ass.*, **80**, 580–619.

Brent, R. P. (1973) *Algorithms for Minimization without Derivatives*, Prentice-Hall, Englewood Cliffs.

Brillinger, D. R. (2001) *Time Series: Data Analysis and Theory*, SIAM, Philadelphia.

Brockwell, P. J. and Davis, R. A. (1991) *Time Series: Theory and Methods*, Springer, New York.

Bruce, A. G. and Gao, H.-Y. (1996) Understanding WaveShrink: variance and bias estimation, *Biometrika*, **83**, 727–745.

Buckheit, J. and Donoho, D. L. (1995) WaveLab and reproducible research, in A. Antoniadis and G. Oppenheim, eds., *Wavelets and Statistics*, volume 103 of *Lecture Notes in Statistics*, pp. 55–82, Springer-Verlag, New York.

Bui, T. D. and Chen, G. (1998) Translation-invariant denoising using multi-wavelets, *IEEE Trans. Sig. Proc*, **46**, 3414–3420.

Burrus, C. S., Gopinath, R. A., and Guo, H. (1997) *Introduction to Wavelets and Wavelet Transforms: A Primer*, Prentice Hall, Upper Saddle River, NJ.

Cai, T. (1999) Adaptive wavelet estimation: a block thresholding and oracle inequality approach, *Ann. Statist.*, **27**, 898–924.

Cai, T. (2002) On block thresholding in wavelet regression: Adaptivity, block size and threshold level, *Statistica Sinica*, **12**, 1241–1273.

Cai, T. and Brown, L. (1998) Wavelet shrinkage for non-equispaced samples, *Ann. Statist.*, **26**, 1783–1799.

Cai, T. and Brown, L. (1999) Wavelet estimation for samples with random uniform design, *Stat. Prob. Lett.*, **42**, 313–321.

Cai, T. and Silverman, B. W. (2001) Incorporating information on neighbouring coefficients into wavelet estimation, *Sankhyā* B, **63**, 127–148.

Cai, T. and Zhou, H. (2008) A data-driven block thresholding approach to wavelet estimation, *Ann. Statist.*, **36**, (to appear).

Candes, E. J. and Donoho, D. L. (2005a) Continuous curvelet transform — I. resolution of the wavefront set, *App. Comp. Harm. Anal.*, **19**, 162–197.

Candes, E. J. and Donoho, D. L. (2005b) Continuous curvelet transform — II. discretization and frames, *App. Comp. Harm. Anal.*, **19**, 198–222.

Cavalier, L. and Raimondo, M. (2007) Wavelet deconvolution with noisy eigenvalues, *IEEE Trans. Sig. Proc*, **55**, 2414–2424.

Chatfield, C. (2003) *The Analysis of Time Series: An Introduction*, Chapman and Hall/CRC, London.

Chesneau, C. (2007) Wavelet block thresholding for samples with random design: a minimax approach under the L^p risk, *Elec. J. Stat.*, **1**, 331–346.

Chiann, C. and Morettin, P. (1999) A wavelet analysis for time series, *J. Nonparam. Statist.*, **10**, 1–46.

Chicken, E. (2003) Block thresholding and wavelet estimation for nonequi-spaced samples, *J. Statist. Plan. Inf.*, **116**, 113–129.

Chicken, E. (2005) Block-dependent thresholding in wavelet regression, *J. Nonparam. Statist.*, **17**, 467–491.

Chicken, E. (2007) Nonparametric regression with sample design following a random process, *Comm. Stat. - Theory and Methods*, **36**, 1915–1934.

Chicken, E. and Cai, T. (2005) Block thresholding for density estimation: local and global adaptivity, *J. Mult. Anal.*, **95**, 76–106.

Chipman, H., Kolaczyk, E., and McCulloch, R. (1997) Adaptive Bayesian wavelet shrinkage, *J. Am. Statist. Ass.*, **92**, 1413–1421.

Chui, C. K. (1997) *Wavelets: a Mathematical Tool for Signal Analysis*, SIAM, Philadelphia.

Chui, C. K. and Lian, J. (1996) A study of orthonormal multi-wavelets, *Applied Numerical Mathematics*, **20**, 273–298.

Claypoole, R. L., Baraniuk, R. G., and Nowak, R. D. (2003) Nonlinear wavelet transforms for image coding via lifting., *IEEE Trans. Im. Proc.*, **12**, 1513–1516.

Clonda, D., Lina, J.-M., and Goulard, B. (2004) Complex Daubechies wavelets: properties and statistical image modelling, *Sig. Proc.*, **84**, 1–23.

Clyde, M. and George, E. I. (2004) Model uncertainty, *Statist. Sci.*, **19**, 81–94.

Clyde, M., Parmigiani, G., and Vidakovic, B. (1998) Multiple shrinkage and subset selection in wavelets, *Biometrika*, **85**, 391–402.

Cohen, A., Daubechies, I., and Feauveau, J. C. (1992) Biorthogonal bases of compactly supported wavelets, *Comm. Pure Appl. Math.*, **45**, 485–500.

Cohen, A., Daubechies, I., and Vial, P. (1993) Wavelets on the interval and fast wavelet transforms, *App. Comp. Harm. Anal.*, **1**, 54–81.

Coifman, R. R. and Donoho, D. L. (1995) Translation-invariant de-noising, in A. Antoniadis and G. Oppenheim, eds., *Wavelets and Statistics*, volume 103 of *Lecture Notes in Statistics*, pp. 125–150, Springer-Verlag, New York.

Coifman, R. R. and Wickerhauser, M. V. (1992) Entropy-based algorithms for best-basis selection, *IEEE Trans. Inf. Th.*, **38**, 713–718.

Dahlhaus, R. (1996) On the Kullback-Leibler information divergence of locally stationary processes, *Stoch. Process. Appl.*, **62**, 139–168.

Dahlhaus, R. (1997) Fitting time series models to nonstationary processes, *Ann. Statist.*, **25**, 1–37.

Dahlhaus, R. and Subba Rao, S. (2006) Statistical inference for time-varying ARCH processes, *Ann. Statist.*, **34**, 1075–1114.

Dahlhaus, R. and Subba Rao, S. (2007) A recursive online algorithm for the estimation of time-varying ARCH parameters, *Bernoulli*, **13**, 389–422.

Daubechies, I. (1988) Orthonormal bases of compactly supported wavelets, *Comms Pure Appl. Math.*, **41**, 909–996.

Daubechies, I. (1992) *Ten Lectures on Wavelets*, SIAM, Philadelphia.

Daubechies, I. and Lagarias, J. C. (1992) Two-scale difference equations II: local regularity, infinite products of matrices and fractals, *SIAM J. Math. Anal.*, **23**, 1031–1079.

Delouille, V., Franke, J., and von Sachs, R. (2004a) Nonparametric stochastic regression with design-adapted wavelets, *Sankhyā* A, **63**, 328–366.

Delouille, V., Simoens, J., and von Sachs, R. (2004b) Smooth design-adapted wavelets for nonparametric stochastic regression, *J. Am. Statist. Ass.*, **99**, 643–658.

Delyon, B. and Juditsky, A. (1996) On minimax wavelet estimators, *App. Comp. Harm. Anal.*, **3**, 215–228.

Donoho, D. L. (1993a) Nonlinear wavelet methods of recovery for signals, densities, and spectra from indirect and noisy data, in *Proceedings of Symposia in Applied Mathematics*, volume 47, American Mathematical Society, Providence: RI.

Donoho, D. L. (1993b) Unconditional bases are optimal bases for data compression and statistical estimation, *App. Comp. Harm. Anal.*, **1**, 100–115.

Donoho, D. L. (1995a) De-noising by soft-thresholding, *IEEE Trans. Inf. Th.*, **41**, 613–627.

Donoho, D. L. (1995b) Nonlinear solution of linear inverse problems by wavelet-vaguelette decomposition, *App. Comp. Harm. Anal.*, **2**, 101–26.

Donoho, D. L. (1999) Wedgelets: Nearly minimax estimation of edges, *Ann. Statist.*, **27**, 859–897.

Donoho, D. L. and Johnstone, I. M. (1994a) Ideal denoising in an orthonormal basis chosen from a library of bases, *Compt. Rend. Acad. Sci. Paris A*, **319**, 1317–1322.

Donoho, D. L. and Johnstone, I. M. (1994b) Ideal spatial adaptation by wavelet shrinkage, *Biometrika*, **81**, 425–455.

Donoho, D. L. and Johnstone, I. M. (1995) Adapting to unknown smoothness via wavelet shrinkage, *J. Am. Statist. Ass.*, **90**, 1200–1224.

Donoho, D. L. and Raimondo, M. (2004) Translation invariant deconvolution in a periodic setting, *Int. J. Wavelets, Multiresolution, Inf. Process.*, **14**, 415–423.

Donoho, D. L., Johnstone, I. M., Kerkyacharian, G., and Picard, D. (1995) Wavelet shrinkage: Asymptopia? (with discussion), *J. R. Statist. Soc. B*, **57**, 301–369.

Donoho, D. L., Johnstone, I. M., Kerkyacharian, G., and Picard, D. (1996) Density estimation by wavelet thresholding, *Ann. Statist.*, **24**, 508–539.

Downie, T. R. (1997) *Wavelets in Statistics*, Ph.D. thesis, University of Bristol, U.K.

Downie, T. R. and Silverman, B. W. (1998) The discrete multiple wavelet transform and thresholding methods, *IEEE Trans. Sig. Proc*, **46**, 2558–2561.

242   References

Eckley, I. A. (2001) *Wavelet Methods for Time Series and Spatial Data*, Ph.D. thesis, University of Bristol, U.K.

Eckley, I. A. and Nason, G. P. (2005) Efficient computation of the discrete autocorrelation wavelet inner product matrix., *Statistics and Computing*, **15**, 83–92.

Eubank, R. L. (1999) *Nonparametric Regression and Spline Smoothing*, Marcel Dekker, New York.

Fadili, M. J., Mathieu, J., and Desvignes, M. (2003) Bayesian wavelet-based Poisson intensity estimation of images using the Fisz transform, in *International conference on image and signal processing*, volume 1, pp. 242–253.

Fan, J. (1996) *Local Polynomial Modelling and Its Applications*, Chapman and Hall, London.

Fisz, M. (1955) The limiting distribution of a function of two independent random variables and its statistical application, *Colloquium Mathematicum*, **3**, 138–146.

Friedman, J. H. and Stuetzle, W. (1981) Projection pursuit regression, *J. Am. Statist. Ass.*, **76**, 817–823.

Fryzlewicz, P. (2003) *Wavelet Techniques for Time Series and Poisson Data*, Ph.D. thesis, University of Bristol, U.K.

Fryzlewicz, P. (2007) Unbalanced Haar technique for nonparametric function estimation, *J. Am. Statist. Ass.*, **102**, 1318–1327.

Fryzlewicz, P. and Delouille, V. (2005) A data-driven Haar-Fisz transform for multiscale variance stabilization, in *Proceedings of the 13th IEEE Workshop on Statistical Signal Processing, Bordeaux*, pp. 539–544.

Fryzlewicz, P. and Nason, G. P. (2003) A Haar-Fisz algorithm for Poisson intensity estimation, Technical Report 03:03, Statistics Group, Department of Mathematics, University of Bristol.

Fryzlewicz, P. and Nason, G. P. (2004) A Haar-Fisz algorithm for Poisson intensity estimation, *J. Comp. Graph. Stat.*, **13**, 621–638.

Fryzlewicz, P. and Nason, G. P. (2006) Haar-Fisz estimation of evolutionary wavelet spectra, *J. R. Statist. Soc. B*, **68**, 611–634.

Fryzlewicz, P., Van Bellegem, S., and von Sachs, R. (2003) Forecasting non-stationary time series by wavelet process modelling, *Ann. Inst. Statist. Math.*, **55**, 737–764.

Fryzlewicz, P., Sapatinas, T., and Subba Rao, S. (2006) A Haar-Fisz technique for locally stationary volatility estimation, *Biometrika*, **93**, 687–704.

Fryzlewicz, P., Delouille, V., and Nason, G. P. (2007) GOES-8 X-ray sensor variance stabilization using the multiscale data-driven Haar-Fisz transform., *J. R. Statist. Soc. C*, **56**, 99–116.

Gabbanini, F., Vannucci, M., Bartoli, G., and Moro, A. (2004) Wavelet packet methods for the analysis of variance of time series with applications to crack widths on the Brunelleschi dome, *J. Comp. Graph. Stat.*, **13**, 639–658.

Gao, H.-Y. (1993) *Wavelet estimation of spectral densities in time series analysis*, Ph.D. thesis, University of California, Berkeley, USA.

Gao, H.-Y. and Bruce, A. G. (1997) WaveShrink with firm shrinkage, *Statistica Sinica*, **4**, 855–874.

Gencay, R., Selcuk, F., and Whitcher, B. (2001) *An Introduction to Wavelets and Other Filtering Methods in Finance and Economics*, Academic Press, San Diego.

Genovese, C. R. and Wasserman, L. (2005) Confidence sets for nonparametric wavelet regression, *Ann. Statist.*, **33**, 698–729.

Geronimo, J. S., Hardin, D. P., and Massopust, P. R. (1994) Fractal functions and wavelet expansions based on several scaling functions, *J. Approx. Theory*, **78**, 373–401.

Ghorai, J. K. and Yu, D. (2004) Data-based resolution selection in positive wavelet density estimation, *Comm. Stat. - Theory and Methods*, **2004**, 2393–2408.

Ghosh, J. K. and Ramamoorthi, R. V. (2003) *Bayesian Nonparametrics*, Springer, New York.

Ghugre, N. R., Martin, M., Scadeng, M., Ruffins, S., Hiltner, T., Pautler, R., Waters, C., Readhead, C., Jacobs, R., and Wood, J. C. (2003) Superiority of 3D wavelet-packet denoising in MR microscopy, *Mag. Res. Imag.*, **21**, 913–921.

Goodman, T. N. T. and Lee, S. L. (1994) Wavelets of multiplicity r, *Trans. Am. Math. Soc.*, **342**, 307–324.

Green, P. J. and Silverman, B. W. (1993) *Nonparametric Regression and Generalized Linear Models*, Chapman and Hall, London.

Haar, A. (1910) Zur theorie der orthogonalen funktionensysteme, *Math. Ann.*, **69**, 331–371.

Hall, P. and Nason, G. P. (1997) On choosing a non-integer resolution level when using wavelet methods, *Stat. Prob. Lett.*, **34**, 5–11.

Hall, P. and Patil, P. (1995) Formulae for mean integrated squared error of nonlinear wavelet-based density estimators, *Ann. Statist.*, **23**, 905–928.

Hall, P. and Patil, P. (1996) On the choice of smoothing parameter, threshold and truncation in nonparametric regression by non-linear wavelet methods, *J. R. Statist. Soc. B*, **58**, 361–377.

Hall, P. and Penev, S. (2001) Cross-validation for choosing resolution level for nonlinear wavelet curve estimators, *Bernoulli*, **7**, 317–341.

Hall, P. and Turlach, B. (1997) Interpolation methods for nonlinear wavelet regression with irregularly spaced design, *Ann. Statist.*, **25**, 1912–1925.

Hall, P. and Wolff, R. C. L. (1995) Estimators of integrals of powers of density derivatives, *Stat. Prob. Lett.*, **24**, 105–110.

Hall, P., Penev, S., Kerkyacharian, G., and Picard, D. (1997) Numerical performance of block thresholded wavelet estimators, *Statistics and Computing*, **7**, 115–124.

Hall, P., Kerkyacharian, G., and Picard, D. (1999) On the minimax optimality of block thresholded wavelet estimators, *Statistica Sinica*, **9**, 33–50.

Hamilton, J. D. (1994) *Time Series Analysis*, Princeton University Press, Princeton, New Jersey.

Hannan, E. J. (1960) *Time series analysis*, Chapman and Hall, London.

Härdle, W. (1992) *Applied Nonparametric Regression*, Cambridge University Press, Cambridge.

Hazewinkel, M. (2002) Orthogonal series, in *Online Encyclopaedia of Mathematics*, Springer, http://eom.springer.de.

Heil, C. and Walnut, D. F. (1989) Continuous and discrete wavelet transforms, *SIAM Rev.*, **31**, 628–666.

Heil, C. and Walnut, D. F. (2006) *Fundamental Papers in Wavelet Theory*, Princeton University Press, Princeton, New Jersey.

Herrick, D. R. M. (2000) *Wavelet Methods for Curve Estimation*, Ph.D. thesis, University of Bristol, U.K.

Herrick, D. R. M., Nason, G. P., and Silverman, B. W. (2001) Some new methods for wavelet density estimation, *Sankhyā* A, **63**, 391–411.

Hess–Nielsen, N. and Wickerhauser, M. V. (1996) Wavelets and time-frequency analysis, *Proceedings of the IEEE*, **84**, 523–540.

Holschneider, M., Kronland–Martinet, R., Morlet, J., and Tchamitchian, P. (1989) A real-time algorithm for signal analysis with the help of the wavelet transform, in J. Combes, A. Grossman, and P. Tchamitchian, eds., *Wavelets, Time-Frequency Methods and Phase Space*, pp. 286–297, Springer, New York.

Houdré, C. and Averkamp, R. (2005) Wavelet thresholding for non necessarily Gaussian noise: Functionality, *Ann. Statist.*, **33**, 2164–2193.

Hsung, T. C., Sun, M. C., Lun, D. P. K., and Siu, W. C. (2003) Symmetric prefilters for multiwavelets, *IEE Proceedings – Vision, image and signal processing*, **150**, 59–68.

Huang, S.-Y. (1997) Wavelet based empirical Bayes estimation for the uniform distribution, *Stat. Prob. Lett.*, **32**, 141–146.

Hunt, K. and Nason, G. P. (2001) Wind speed modelling and short-term prediction using wavelets, *Wind Engineering*, **25**, 55–61.

Jansen, M. (2001) *Noise Reduction by Wavelet Thresholding*, Springer, New York.

Jansen, M. (2006) Multiscale Poisson data smoothing, *J. R. Statist. Soc. B*, **68**, 27–48.

Jansen, M. and Bultheel, A. (1999) Multiple wavelet threshold estimation by generalized cross-validation for images with correlated noise, *IEEE Trans. Im. Proc.*, **8**, 947–953.

Jansen, M. and Oonincx, P. (2005) *Second generation wavelets and applications*, Springer Verlag, Berlin.

Jansen, M., Malfait, M., and Bultheel, A. (1997) Generalized cross validation for wavelet thresholding, *Sig. Proc.*, **56**, 33–44.

Jansen, M., Nason, G. P., and Silverman, B. W. (2001) Scattered data smoothing by empirical Bayesian shrinkage of second generation wavelet coefficients, in M. Unser and A. Aldroubi, eds., *Wavelet applications in signal and image processing. Proceedings of SPIE*, volume 4478, pp. 87–97.

Jawerth, B. and Sweldens, W. (1994) An overview of wavelet based multiresolution analysis, *SIAM Rev.*, **36**, 377–412.

Johnson, N. L. (1949) Systems of frequency curves generated by methods of translation, *Biometrika*, **36**, 149–176.

Johnstone, I. M. and Silverman, B. W. (1990) Speed of estimation in positron emission tomography and related inverse problems, *Ann. Statist.*, **18**, 251–80.

Johnstone, I. M. and Silverman, B. W. (1991) Discretization effects in statistical inverse problems, *J. Complex.*, **7**, 1–34.

Johnstone, I. M. and Silverman, B. W. (1997) Wavelet threshold estimators for data with correlated noise, *J. R. Statist. Soc.* B, **59**, 319–351.

Johnstone, I. M. and Silverman, B. W. (2004) Needles and hay in haystacks: Empirical Bayes estimates of possibly sparse sequences, *Ann. Statist.*, **32**, 1594–1649.

Johnstone, I. M. and Silverman, B. W. (2005a) Ebayesthresh: R programs for empirical Bayes thresholding, *J. Statist. Soft.*, **12**, 1–38.

Johnstone, I. M. and Silverman, B. W. (2005b) Empirical Bayes selection of wavelet thresholds, *Ann. Statist.*, **33**, 1700–1752.

Johnstone, I. M., Kerkyacharian, G., and Picard, D. (1992) Density estimation using wavelet methods, *Compt. Rend. Acad. Sci. Paris I*, **315**, 211–216.

Johnstone, I. M., Kerkyacharian, G., Picard, D., and Raimondo, M. (2004) Wavelet deconvolution in a periodic setting (with discussion), *J. R. Statist. Soc.* B, **66**, 547–573.

Juditsky, A. and Lambert–Lacroix, S. (2004) On minimax density estimation on $\mathbb{R}$, *Bernoulli*, **10**, 187–220.

Kato, T. (1999) Density estimation by truncated wavelet expansion, *Stat. Prob. Lett.*, **43**, 159–168.

Kerkyacharian, G. and Picard, D. (1992a) Density estimation by the kernel and wavelet methods – link between kernel geometry and regularity constraints, *Compt. Rend. Acad. Sci. Paris I*, **315**, 79–84.

Kerkyacharian, G. and Picard, D. (1992b) Density estimation in Besov spaces, *Stat. Prob. Lett.*, **13**, 15–24.

Kerkyacharian, G. and Picard, D. (1993) Density estimation by kernel and wavelets methods — optimality of Besov spaces, *Stat. Prob. Lett.*, **18**, 327–336.

Kerkyacharian, G. and Picard, D. (1997) Limit of the quadratic risk in density estimation using linear methods, *Stat. Prob. Lett.*, **31**, 299–312.

Kim, D. and Oh, H.-S. (2006) CVThresh: R package for level-dependent cross-validation thresholding, *J. Statist. Soft.*, **15**.

Kingman, J. F. C. and Taylor, S. C. (1966) *Introduction to Measure and Probability*, Cambridge University Press, Cambridge.

Knight, M. and Nason, G. P. (2008) A nondecimated lifting transform, *Statistics and Computing*, **18**, (to appear).

Kolaczyk, E. D. (1994) *Wavelet Methods for the Inversion of Certain Homogeneous Linear Operators in the Presence of Noisy Data*, Ph.D. thesis, Department of Statistics, Stanford University, Stanford, CA, USA.

Kolaczyk, E. D. (1996) A wavelet shrinkage approach to tomographic image reconstruction, *J. Am. Statist. Ass.*, **91**, 1079–1090.

Kolaczyk, E. D. (1997) Non-parametric estimation of Gamma-ray burst intensities using Haar wavelets, *The Astrophysical Journal*, **483**, 340–349.

Kolaczyk, E. D. (1999a) Bayesian multiscale models for Poisson processes, *J. Am. Statist. Ass.*, **94**, 920–933.

Kolaczyk, E. D. (1999b) Wavelet shrinkage estimation of certain Poisson intensity signals using corrected thresholds, *Statistica Sinica*, **9**, 119–135.

Koo, J. Y. and Kim, W. C. (1996) Wavelet density estimation by approximation of log-densities, *Stat. Prob. Lett.*, **26**, 271–278.

Kovac, A. and Silverman, B. W. (2000) Extending the scope of wavelet regression methods by coefficient-dependent thresholding, *J. Am. Statist. Ass.*, **95**, 172–183.

Kovačević, J. and Vetterli, M. (1992) Nonseparable multidimensional perfect reconstruction filter banks and wavelet bases for $\mathbb{R}^n$, *IEEE Trans. Inf. Th.*, **38**, 533–555.

Lawton, W. (1993) Applications of complex valued wavelet transforms to subband decomposition, *IEEE Trans. Sig. Proc*, **41**, 3566–3568.

Le Pennec, E. and Mallat, S. G. (2005a) Bandelet image approximation and compression, *Multiscale Model. and Simul.*, **4**, 992–1039.

Le Pennec, E. and Mallat, S. G. (2005b) Sparse geometric image representations with bandelets, *IEEE Trans. Im. Proc.*, **14**, 423–438.

Leblanc, F. (1993) Density estimation using wavelet methods, *Compt. Rend. Acad. Sci. Paris I*, **317**, 201–204.

Leblanc, F. (1995) Wavelet density-estimation of a continuous-time process and application to diffusion process, *Compt. Rend. Acad. Sci. Paris I*, **321**, 345–350.

Leblanc, F. (1996) Wavelet linear density estimator for a discrete-time stochastic process: l_p-losses, *Stat. Prob. Lett.*, **27**, 71–84.

Li, L.-Y. (2002) Hazard rate estimation for censored data by wavelet methods, *Comm. Stat. - Theory and Methods*, **31**, 943–960.

Li, T.-H. (1999) Multiscale representation and analysis of spherical data by spherical wavelets, *SIAM J. Sci. Comput.*, **21**, 924–953.

Li, Y.-Z. (2005) On the construction of a class of bidimensional nonseparable compactly supported wavelets, *Proc. Am. Math. Soc.*, **133**, 3505–3513.

Liang, H.-Y., Mammitzsch, V., and Steinebach, J. (2005) Nonlinear wavelet density and hazard rate estimation for censored data under dependent observations, *Statistics and Decisions*, **23**, 161–180.

Lina, J.-M. (1997) Image processing with complex Daubechies wavelets, *Journal of Mathematical Imaging and Vision*, **7**, 211–223.

Lina, J.-M. and MacGibbon, B. (1997) Non-linear shrinkage estimators with complex Daubechies wavelets, in *Proceedings of SPIE*, volume 3169, pp. 67–79.

Lina, J.-M. and Mayrand, M. (1995) Complex Daubechies wavelets, *App. Comp. Harm. Anal.*, **2**, 219–229.

Lina, J.-M., Turcotte, P., and Goulard, B. (1999) Complex dyadic multiresolution analysis, in *Advances in Imaging and Electron Physics*, volume 109, Academic Press.

Mallat, S. G. (1989a) Multiresolution approximations and wavelet orthonormal bases of $L^2(R)$, *Trans. Am. Math. Soc.*, **315**, 69–87.

Mallat, S. G. (1989b) A theory for multiresolution signal decomposition: the wavelet representation, *IEEE Trans. Patt. Anal. and Mach. Intell.*, **11**, 674–693.

Mallat, S. G. (1991) Zero-crossings of a wavelet transform, *IEEE Trans. Inf. Th.*, **37**, 1019–1033.

Mallat, S. G. (1998) *A Wavelet Tour of Signal Processing*, Academic Press, San Diego.

Mallat, S. G. and Hwang, W. L. (1992) Singularity detection and processing with wavelets, *IEEE Trans. Inf. Th.*, **38**, 617–643.

Mallat, S. G. and Zhang, Z. (1993) Matching pursuits with time-frequency dictionaries, *IEEE Trans. Sig. Proc*, **41**, 3397–3415.

Mardia, K. V., Kent, J. T., and Bibby, J. M. (1979) *Multivariate Analysis*, Academic Press, New York.

Marron, J. S. and Wand, M. P. (1992) Exact mean integrated squared error, *Ann. Statist.*, **20**, 712–736.

Masry, E. (1994) Probability density estimation from dependent observations using wavelet orthonormal bases, *Stat. Prob. Lett.*, **21**, 181–194.

Masry, E. (1997) Multivariate probability density estimation by wavelet methods: Strong consistency and rates for stationary time series, *Stoch. Process. Appl.*, **67**, 177–193.

Meyer, Y. (1993a) *Wavelets: Algorithms and Applications*, SIAM, Philadelphia.

Meyer, Y. (1993b) *Wavelets and Operators*, Cambridge University Press, Cambridge.

Morlet, J., Arens, G., Fourgeau, E., and Giard, D. (1982) Wave propagation and sampling theory — Part I: complex signal and scattering in multilayered media, *Geophysics*, **47**, 203–221.

Motakis, E. S., Nason, G. P., Fryzlewicz, P., and Rutter, G. A. (2006) Variance stabilization and normalization for one-color microarray data using a data-driven multiscale approach, *Bioinformatics*, **22**, 2547–2553.

Muller, P. and Vidakovic, B. (1999) Bayesian inference with wavelets: density estimation, *J. Comp. Graph. Stat.*, **7**, 456–468.

Nason, G. P. (1996) Wavelet shrinkage using cross-validation, *J. R. Statist. Soc. B*, **58**, 463–479.

Nason, G. P. (2002) Choice of wavelet smoothness, primary resolution and threshold in wavelet shrinkage, *Statistics and Computing*, **12**, 219–227.

Nason, G. P. (2006) Stationary and non-stationary time series, in H. Mader and S. Coles, eds., *Statistics in Volcanology*, pp. 129–142, Geological Society of London, London.

Nason, G. P. and Bailey, D. (2008) Estimating the intensity of conflict in Iraq, *J. R. Statist. Soc.* A, **171**, (to appear).

Nason, G. P. and Sapatinas, T. (2002) Wavelet packet transfer function modelling of nonstationary time series, *Statistics and Computing*, **12**, 19–56.

Nason, G. P. and Silverman, B. W. (1994) The discrete wavelet transform in S, *J. Comp. Graph. Stat.*, **3**, 163–191.

Nason, G. P. and Silverman, B. W. (1995) The stationary wavelet transform and some statistical applications, in A. Antoniadis and G. Oppenheim, eds., *Wavelets and Statistics*, volume 103 of *Lecture Notes in Statistics*, pp. 281–229, Springer-Verlag, New York.

Nason, G. P. and von Sachs, R. (1999) Wavelets in time series analysis, *Phil. Trans. R. Soc. Lond.* A, **357**, 2511–2526.

Nason, G. P., von Sachs, R., and Kroisandt, G. (2000) Wavelet processes and adaptive estimation of the evolutionary wavelet spectrum, *J. R. Statist. Soc.* B, **62**, 271–292.

Nason, G. P., Sapatinas, T., and Sawczenko, A. (2001) Wavelet packet modelling of infant sleep state using heart rate data, *Sankhyā* B, **63**, 199–217.

Neelamani, R., Choi, H., and Baraniuk, R. (2004) ForWaRD: Fourier-wavelet regularized deconvolution for ill-conditioned systems, *IEEE Trans. Sig. Proc*, **52**, 418–433.

Neumann, M. and von Sachs, R. (1995) Wavelet thresholding: beyond the Gaussian iid situation, in A. Antoniadis and G. Oppenheim, eds., *Wavelets and Statistics*, volume 103 of *Lecture Notes in Statistics*, Springer-Verlag, New York.

Nguyen, T. Q. and Vaidyanathan, P. P. (1989) Two-channel perfect-reconstruction fit QMF structures which yield linear-phase analysis and synthesis filters., *IEEE Trans. Acoust., Speech*, **37**, 676–690.

Nowak, R. D. and Baraniuk, R. G. (1999) Wavelet domain filtering for photon imaging systems, *IEEE Trans. Im. Proc.*, **8**, 666–678.

Nunes, M. (2006) *Some New Multiscale Methods for Curve Estimation and Binomial Data*, Ph.D. thesis, University of Bristol, U.K.

Nunes, M., Knight, M., and Nason, G. P. (2006) Adaptive lifting for nonparametric regression, *Statistics and Computing*, **16**, 143–159.

Ogden, R. T. (1997) *Essential Wavelets for Statistical Applications and Data Analysis*, Birkhauser, Boston.

Oh, H.-S. (1999) *Spherical Wavelets and their Statistical Analysis with Applications to Meterological Data.*, Ph.D. thesis, Department of Statistics, Texas A&M University, College Station, Texas, USA.

Oh, H.-S. and Li, T.-H. (2004) Estimation of global temperature fields from scattered observations by a spherical-wavelet-based spatially adaptive method, *J. R. Statist. Soc. B*, **66**, 221–238.

O'Hagan, A. and Forster, J. (2004) *Bayesian Inference*, Arnold, London.

Ombao, H., von Sachs, R., and Guo, W. S. (2005) SLEX analysis of multivariate nonstationary time series, *J. Am. Statist. Ass.*, **100**, 519–531.

Ombao, H. C., Raz, J., von Sachs, R., and Malow, B. A. (2001) Automatic statistical analysis of bivariate nonstationary time series, *J. Am. Statist. Ass.*, **96**, 543–560.

Ombao, H. C., Raz, J., von Sachs, R., and Guo, W. (2002) The SLEX model of non-stationary random processes, *Ann. Inst. Statist. Math.*, **54**, 171–200.

O'Sullivan, F. (1986) A statistical perspective on ill-posed inverse problems (with discussion), *Statist. Sci.*, **1**, 502–27.

Patil, P. (1997) Nonparametric hazard rate estimation by orthogonal wavelet methods., *J. Statist. Plan. Inf.*, **60**, 153–168.

Pensky, M. and Vidakovic, B. (2001) On non-equally spaced wavelet regression, *Ann. Inst. Statist. Math.*, **53**, 681–690.

Percival, D. B. (1995) On estimation of the wavelet variance, *Biometrika*, **82**, 619–631.

Percival, D. B. and Guttorp, P. (1994) Long-memory processes, the Allan variance and wavelets., in E. Foufoula-Georgiou and P. Kumar, eds., *Wavelets in Geophysics*, pp. 325–57, Academic Press, New York.

Percival, D. B. and Walden, A. T. (2000) *Wavelet Methods for Time Series Analysis*, Cambridge University Press, Cambridge.

Pesquet, J.-C., Krim, H., and Carfantan, H. (1996) Time-invariant orthonormal wavelet representations, *IEEE Trans. Sig. Proc*, **44**, 1964–1970.

Picard, D. and Tribouley, K. (2000) Adaptive confidence interval for pointwise curve estimation, *Ann. Statist.*, **28**, 298–335.

Picklands, J. (1967) Maxima of stationary Gaussian processes, *Probability Theory and Related Fields*, **7**, 190–223.

Piessens, R., de Doncker-Kapenga, E., Uberhuber, C. W., and Kahaner, D. K. (1983) *QUADPACK: a subroutine package for automatic integration.*, Springer-Verlag, New York.

Pinheiro, A. and Vidakovic, B. (1997) Estimating the square root of a density via compactly supported wavelets., *Comp. Stat. Data Anal.*, **25**, 399–415.

Press, W. H., Teukolsky, S. A., Vetterling, W. T., and Flannery, B. P. (1992) *Numerical Recipes in C, the Art of Scientific Computing*, Cambridge University Press, Cambridge.

Priestley, M. B. (1965) Evolutionary spectra and non-stationary processes, *J. Roy. Stat. Soc. Series B*, **27**, 204–237.

Priestley, M. B. (1983) *Spectral Analysis and Time Series*, Academic Press, London.

Priestley, M. B. and Subba Rao, T. (1969) A test for non-stationarity of time-series, *J. R. Statist. Soc. B*, **31**, 140–149.

R Development Core Team (2008) *R: A Language and Environment for Statistical Computing*, R Foundation for Statistical Computing, Vienna, Austria, ISBN 3-900051-07-0.

Raimondo, M. and Stewart, M. (2007) The WaveD transform in R: performs fast translation-invariant wavelet deconvolution, *J. Statist. Soft.*, **21**, 1–23.

Ramlau–Hansen, H. (1983) Smoothing counting processes by means of kernel functions, *Ann. Statist.*, **11**, 453–466.

Renaud, O. (2002a) The discrimination power of projection pursuit with different density estimators, *Biometrika*, **89**, 129–143.

Renaud, O. (2002b) Sensitivity and other properties of wavelet regression and density estimators, *Statist. Sci.*, **12**, 1275–1290.

Rodríguez–Casal, A. and De Uña–Álvarez, J. (2004) Nonlinear wavelet density estimation under the Koziol-Green model, *J. Nonparam. Statist.*, **16**, 91–109.

Romberg, J., Wakin, M., and Baraniuk, R. (2003) Approximation and Compression of Piecewise Smooth Images Using a Wavelet/Wedgelet Geometric Model, in *IEEE International Conference on Image Processing*.

Rong-Qing, J., Riemenschneider, S. D., and Zhou, D.-X. (1998) Vector subdivision schemes and multiple wavelets, *Mathematics of Computation*, **67**, 1533–1563.

Rosenberg, T. (1999) The unfinished revolution of 1989, *Foreign Policy*, **115**, 90–105.

Safavi, A. A., Chen, J., and Romagnoil, J. A. (1997) Wavelet-based density estimation and application to process monitoring, *AICHE Journal*, **43**, 1227–1241.

Sardy, S. (2000) Minimax threshold for denoising complex signals with Waveshrink, *IEEE Trans. Sig. Proc*, **48**, 1023–1028.

Sardy, S., Percival, D. B., Bruce, A. G., Gao, H.-Y., and Stuetzle, W. (1999) Wavelet shrinkage for unequally spaced data, *Statistics and Computing*, **9**, 65–75.

Semandeni, C., Davison, A. C., and Hinkley, D. V. (2004) Posterior probability intervals in Bayesian wavelet estimation, *Biometrika*, **91**, 497–505.

Shensa, M. J. (1992) The discrete wavelet transform: Wedding the à trous and mallat algorithms, *IEEE Trans. Sig. Proc*, **40**, 2464–2482.

Shensa, M. J. (1996) Discrete inverses for nonorthogonal wavelet transforms, *IEEE Trans. Sig. Proc*, **44**, 798–807.

Silverman, B. W. (1985) Some aspects of the spline smoothing approach to nonparametric curve fitting, *J. R. Statist. Soc. B*, **47**, 1–52.

Silverman, B. W. (1986) *Density Estimation for Statistics and Data Analysis*, Chapman and Hall, London.

Silverman, R. A. (1957) Locally stationary random processes, *IRE Trans. Information Theory*, **IT-3**, 182–187.

Simonoff, J. S. (1998) *Smoothing Methods in Statistics*, Springer-Verlag, New York.

Stein, C. (1981) Estimation of the mean of a multivariate normal distribution, *Ann. Statist.*, **9**, 1135–1151.

Stone, M. (1974) Cross-validatory choice and assessment of statistical predictions, *J. R. Statist. Soc. B*, **36**, 111–147.

Strang, G. and Nguyen, T. (1996) *Wavelets and Filter Banks*, Wellesley, MA.

Strang, G. and Strela, V. (1994) Orthogonal multiwavelets with vanishing moments, *Optical Engineering*, **33**, 2104–2107.

Strang, G. and Strela, V. (1995) Short wavelets and matrix dilation equations, *IEEE Trans. Sig. Proc*, **43**, 108–115.

Strela, V., Heller, P. N., Strang, G., Topiwala, P., and Heil, C. (1999) The application of multiwavelet filterbanks to image processing, *IEEE Trans. Im. Proc.*, **8**, 548–563.

Stuart, A. and Ord, J. K. (1994) *Kendall's Advanced Theory of Statistics: Distribution Theory*, volume 1, Arnold, London.

Sweldens, W. (1996) Wavelets and the lifting scheme: a 5 minute tour, *Z. Angew. Math. Mech.*, **76**, 41–44.

Sweldens, W. (1997) The lifting scheme: a construction of second generation wavelets, *SIAM J. Math. Anal.*, **29**, 511–546.

Taubman, D. S. and Marcellin, M. W. (2001) *JPEG2000: image compression fundamentals, standards and practice*, Kluwer, Norwell, Mass.

Tibshirani, R. (1988) Estimating transformations for regression via additivity and variance stabilization, *J. Am. Statist. Ass.*, **83**, 394–405.

Tikhonov, A. (1963) Solution of incorrectly formulated problems and the regularization method, *Soviet Math. Dokl.*, **5**, 1035–1038.

Timmermann, K. E. and Nowak, R. D. (1999) Multiscale modeling and estimation of Poisson processes with application to photon-limited imaging., *IEEE Trans. Inf. Th.*, **45**, 846–862.

Torrence, C. and Compo, G. P. (1998) A practical guide to wavelet analysis, *Bulletin of the American Meteorological Society*, **79**, 61–78.

Vaidyanathan, P. P. (1990) Multirate digital filters, filter banks, polyphase networks and applications: a tutorial, *Proceedings of the IEEE*, **78**, 56–93.

Van Bellegem, S. and von Sachs, R. (2008) Locally adaptive estimation of evolutionary wavelet spectra, *Ann. Statist.*, **36**, (to appear).

Vetterli, M. and Herley, C. (1992) Wavelets and filter banks: Theory and design, *IEEE Trans. Sig. Proc*, **40**, 2207–2232.

Vidakovic, B. (1998) Nonlinear wavelet shrinkage with Bayes rules and Bayes factors, *J. Am. Statist. Ass.*, **93**, 173–179.

Vidakovic, B. (1999a) *Statistical Modeling by Wavelets*, Wiley, New York.

Vidakovic, B. (1999b) Wavelet-based nonparametric Bayes methods, in P. Müller and B. Vidakovic, eds., *Bayesian Inference in Wavelet Based Models*, volume 141 of *Lecture Notes in Statistics*, Springer-Verlag, New York.

von Sachs, R. and MacGibbon, B. (2000) Non-parametric curve estimation by wavelet thresholding with locally stationary errors, *Scand. J. Stat.*, **27**, 475–499.

von Sachs, R. and Neumann, M. H. (2000) A wavelet-based test for station-arity, *J. Time Ser. Anal.*, **21**, 597–613.

Wainwright, M. J., Simoncelli, E. P., and Willsky, A. S. (2001) Random cascades on wavelet trees and their use in analyzing and modeling natural images, *App. Comp. Harm. Anal.*, **11**, 89–123.

Walker, J. S. (2004) Fourier series, in R. Meyers, ed., *Encyclopedia of Physical Sciences and Technology*, pp. 167–183, Academic Press.

Walter, G. G. and Shen, X.-P. (2001) *Wavelets and Other Orthogonal Systems*, Chapman and Hall, Boca Raton.

Walter, G. G. and Shen, X.-P. (2005) Wavelet like behaviour of Slepian functions and their use in density estimation, *Comm. Stat. - Theory and Methods*, **34**, 687–711.

Wand, M. P. and Jones, M. C. (1994) *Kernel Smoothing*, Chapman and Hall, London.

Wang, Y. (1996) Function estimation via wavelet shrinkage for long-memory data, *Ann. Statist.*, **24**, 466–484.

Wasserman, L. (2005) *All of Nonparametric Statistics*, Springer, New York.

West, M. and Harrison, P. (1997) *Bayesian Forecasting and Dynamic Models*, Springer, New York.

Westheimer, G. (2001) The Fourier theory of vision, *Perception*, **30**, 531–541.

Weyrich, N. and Warhola, G. T. (1998) Wavelet shrinkage and generalized cross validation for image denoising, *IEEE Trans. Im. Proc.*, **7**, 82–90.

Wickerhauser, M. V. (1994) *Adapted Wavelet Analysis from Theory to Software*, A.K. Peters, Wellesley, MA.

Williams, D. (1991) *Probability with Martingales*, Cambridge University Press, Cambridge.

Wu, S. and Wells, M. (2003) Nonparametric estimation of hazard functions by wavelet methods, *J. Nonparam. Statist.*, **15**, 187–203.

Xia, X.-G., Geronimo, J., Hardin, D., and Suter, B. (1996) Design of prefilters for discrete multiwavelet transforms, *IEEE Trans. Sig. Proc*, **44**, 25–35.

Yates, F. (1937) The design and analysis of factorial experiments, *Imp. Bur. Soil. Sci. Tech. Comm.*, **35**.

Zaroubi, S. and Goelman, G. (2000) Complex denoising of MR data via wavelet analysis: application for functional MRI, *Mag. Res. Imag.*, **18**, 59–68.

Zhang, B., Fadili, M. J., and Starck, J.-L. (2008) Wavelets, ridgelets and curvelets for Poisson noise removal, *IEEE Trans. Im. Proc.*, **17**, (to appear).

Zhang, C.-H. (2005) General empirical Bayes wavelet methods and adaptive minimax estimation, *Ann. Statist.*, **33**, 54–100.

Zhang, S. and Zheng, Z. (1999) On the asymptotic normality for L_2-error of wavelet density estimator with application, *Comm. Stat. - Theory and Methods*, **28**, 1093–1104.

Index

Multiscale Modeling
A Bayesian Perspective

Marco A.R. Ferreira and Herbert K.H. Lee

The book is aimed at statisticians, applied mathematicians, and engineers working on problems dealing with multiscale processes in time and/or space, such as in engineering, finance, and environmetrics. The book will also be of interest to those working on multiscale computation research. The main prerequisites are knowledge of Bayesian statistics and basic Markov chain Monte Carlo methods. A number of real-world examples are thoroughly analyzed in order to demonstrate the methods and to assist the readers in applying these methods to their own work. To further assist readers, the authors are making source code (for R) available for many of the basic methods discussed herein

2007. 252 pp. (Springer Series in Statistics) Hardcover
ISBN 978-0-387-70897-3

Lattice:
Multivariate Data Visualization with R

Deepayan Sarkar

The book contains close to 150 figures produced with lattice. Many of the examples emphasize principles of good graphical design; almost all use real data sets that are publicly available in various R packages. All code and figures in the book are also available online, along with supplementary material covering more advanced topics.

2008. Approx. 290 pp. (Use R!) Softcover
ISBN 978-0-387-75968-5

Analysis of Integrated and Co-integrated Time Series with R, Second Edition

Bernhard Pfaff

This book encompasses seasonal unit roots, fractional integration, coping with structural breaks, and multivariate time series models. The book is enriched by numerous programming examples to artificial and real data so that it is ideally suited as an accompanying text book to computer lab classes.The second edition adds a discussion of vector autoregressive, structural vector autoregressive, and structural vector error-correction models.

2008. Approx 208 pp. (Use R!) Softcover
ISBN 978-0-387-75966-1

Printed in the United States